THE METHOD OF SCIENCE

A Course in Understanding Science,
based upon the
De Magnete of William Gilbert,
and
the *Vegetable Staticks* of Stephen Hales

THE WYKEHAM SCIENCE SERIES

for schools and universities

To broaden the outlook of the senior grammar school pupil and to introduce the undergraduate to the present state of science as a university study is the aim of the Wykeham Science Series. Each book seeks to reinforce this link between school and university levels, and the main author, a university teacher distinguished in the field, is assisted by an experienced sixth-form schoolmaster.

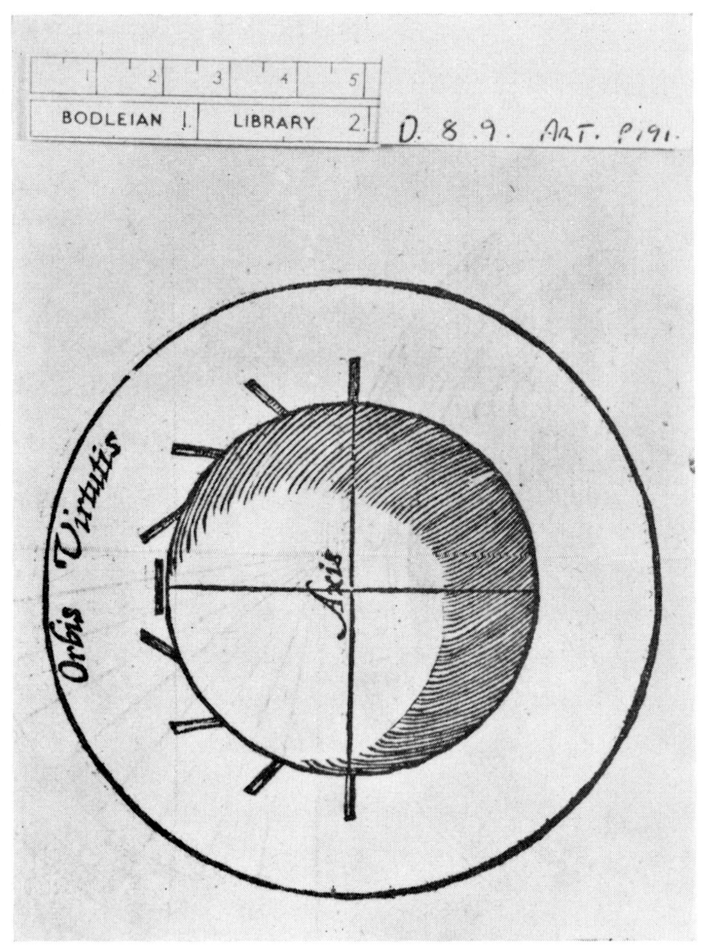

The basic experiment on dip, from the *De Magnete*

THE METHOD OF SCIENCE

R. Harré – Linacre College, Oxford

WYKEHAM PUBLICATIONS (LONDON) LTD
(A subsidiary of Taylor & Francis Ltd)
LONDON & WINCHESTER
1970

First published 1970 by Wykeham Publications (London) Ltd.

Cover illustration: an early Armed Lodestone, by courtesy of the Curator, Museum of the History of Science, Oxford.

Printed in Great Britain by Taylor & Francis Ltd.
10–14 Macklin Street, London, W.C.2

ISBN 0 85109 090 7

Distribution:

UNITED KINGDOM, EUROPE, MIDDLE EAST AND AFRICA

Chapman & Hall Ltd. (a member of Associated Book Publishers Ltd.), 11 New Fetter Lane, London, E.C.4 and North Way, Andover, Hampshire.

UNITED STATES OF AMERICA, CANADA AND MEXICO

Springer-Verlag New York Inc., 175 Fifth Avenue, New York, New York 10010.

AUSTRALIA AND NEW GUINEA

Hicks Smith & Sons Pty., Ltd., 301 Kent Street, Sydney, N.S.W. 2000.

NEW ZEALAND AND FIJI

Hicks Smith & Sons Ltd., 238 Wakefield Street, Wellington.

ALL OTHER TERRITORIES

Taylor & Francis Ltd., 10–14 Macklin Street, London, W.C.2.

PREFACE

THE History and Philosophy of Science are usually presented around those developments in mathematical physics and astronomy which culminated in the work of Sir Isaac Newton, and those biological studies that centre round the theory of Darwin. This book emphasizes the achievements of different men and looks at science in a different way. The experimental tradition is emphasized, and magnetism is taken as the central problem around which the growth of physics is sketched. Instead of the evolution of creatures, their physiology and mode of life are brought out in looking at biology. It is hoped that this will give a new interest to those who have had some of the traditional scientific education in the lower forms at school and help science specialists to get some idea of the way their speciality developed. I hope that this approach may also catch the interest of those who have never before become acquainted with the scientific view of the world in which they live and the study of both its common and its more puzzling phenomena. It is hoped that this book will be studied in class in conjunction with the *De Magnete* and the *Vegetable Staticks**, and that these noble classics of scientific work will be read, and the reasonings and puzzlements and mistakes of their authors followed. It is my belief that the best way to come to understand science is to follow an investigation made by one of the great explorers of nature, when that man tells us what he did and why he did it, and lets us see his mind at work. We follow him in his thoughts and in his practical studies. For this reason there have been many suggestions for experiments included, but the attentive teacher can find many others described in these great books by Gilbert and Hales. In the quotations from the *De Magnete* and the *Vegetable Staticks*, spelling has been modernized.

* Both are now available in paper back, the *De Magnete* from Dover Books, and *Vegetable Staticks* from Oldbourne.

CONTENTS

THE MAGNES OR LODESTONE'S CHALLENGE

by

Robert Norman

Give place ye glittering sparks,
  ye glimmering sapphires bright,
Ye Rubies red, and Diamonds brave,
  wherein ye most delight.

In brief, ye stones enriched
  and burnished all with gold,
Set forth in lapidaries' shops,
  for Jewels to be sold.

Give place, give place I say,
  your beauty, gleam and glee,
Is all the virtue for the which,
  accepted so you be.

Magnes, the Lodestone I,
  your painted sheaths defy,
Without my help in Indian sea,
  the best of you might lie.

I guide the Pilot's course,
  his helping hand I am,
The Mariner delights in me,
  so doth the Merchant man.

My virtue lies unknown,
  my secrets hidden are,
By me the Court and Commonweal,
  are pleasured very far.

No ship could sail on seas,
  her course to run aright,
Nor Compass show the ready way,
  were Magnes not of might.

Blush then and blemish all,
  Bequeath to me that's due,
Your seats in gold, your price in plate,
  which Jewellers do renew.

Its I, its I alone,
  whom you usurp upon,
Magnes by name, the Lodestone called,
  the Prince of Stones alone.

If this you can deny,
  then seem to make reply,
And let the painful seaman judge,
  the which of us doth lie.

The Mariner's Judgement.

The Lodestone is the Stone,
  the only stone alone,
Deserving praise above the rest,
  whose virtues are unknown.

The Merchant's Judgement.

The Sapphires bright, the Diamonds brave,
  are stones that bear the name,
But flatter not, and tell the truth,
  Magnes deserves the fame.

Robert Norman was an instrument maker, and in 1581 discovered the phenomenon of dip. See p. 28.

CHAPTER 1

science before Copernicus

In this book we will trace some of the history of modern science, and try to discover some of its methods. To do this we will look at the work of two great scientists of the early modern period, William Gilbert and Stephen Hales. To understand their greatness we will have to look not only at their work, but at the development of science up to their time, and at some of the subsequent developments of the ideas and methods of research that they initiated. William Gilbert lived from 1546 until 1603, and Stephen Hales from 1677 until 1761. Gilbert studied magnetism and electricity, and Hales studied botany, particularly plant physiology. Each man left the science he studied profoundly changed by his work, both as to the knowledge of fact, and to the methods of investigation brought to bear upon the subject matter. In the first part of this book we will look at some of the influences which led to the development of the physical sciences in the sixteenth and seventeenth centuries, and then proceed to a careful study of Gilbert's own work, repeating if possible his great experiments. We will follow him in his thinking, particularly, as he devises experiments and assesses their significance. In the second part we will do the same kind of study of the origins of plant physiology, and of the particular work of Hales. One reason for choosing these two men for particular study is the candour of their writing, so that to a greater extent than usual we can " follow their thoughts ".

It would be very wrong to think that modern science suddenly came into being with Copernicus's idea that the sun should be chosen as the centre of the astronomical system. The origins of modern science are to be found in the ancient world, and in the use of the ancient methods of scientific research by some of the scientists of the Middle Ages. Modern science gradually developed out of that part of ancient science which had been revived in the Middle Ages, and indeed developed in certain ways. There was no sudden discovery of the scientific method. Here and there in Europe and the Middle East groups of people pursued recognizably scientific studies, many others copied and commented upon the scientific works of the ancients, with varying degrees of criticism. Some sort of scientific education was gained by a steadily increasing number of people. At the same time there were other kinds of development. The two centuries preceding the great days of Descartes, Boyle, Newton and their contemporaries were a time of great interest in magic, and full of attempts to make magic work. It is not too much to say that even Newton was in some ways an adept, perhaps the last great magician, as well as the first great scientist of the modern era. We shall see that his ideas about gravity stem in part from the magical tradition. There were important developments in engineering and technology, too, in the period immediately preceding the Scientific Revolution, and from these, new instruments as well as new principles of action were derived. The development of the clock alone was of profound importance in the origin of modern science. We shall touch upon some of these strands as we look for the beginnings of the sciences of magnetism and of plant physiology in medieval times. The science of the Middle Ages was Aristotelian science, to a very large degree. We shall begin by finding out what the Aristotelian method was, and look at some fine scientific research which was done by following that method. Aristotle himself discusses scientific method in several places in his extensive works. We shall draw upon two of these for our account, the *Metaphysics* and the *Posterior Analytics*. You can get quite a good idea of the way the Greeks thought about the world by reading Book A of the *Metaphysics*. The *Posterior Analytics* is not so readily followed as the discussion in the *Metaphysics*, but you might have a glance at Book II of the *Posterior Analytics* from which later scientists took their ideas of Aristotle's method.

Aristotle begins his account of science by pointing out that there are two kinds of knowledge. We can state his line of argument as follows. We can know *that* the kettle is boiling, that the door cannot be opened, that sodium reacts with water, that the Earth moves in an elliptical orbit around the sun, that plants without an adequate supply of water die, that a piece of glass rubbed with silk will attract small scraps of paper. But we may also enquire as to *why* these facts are so, and so open up a different realm of knowledge. It is to know something more about the situation to know why the kettle is boiling in addition to knowing that it is. The same is true of the other cases we have mentioned. To know why sodium reacts with water is to know something more than that it does. In each case the extra knowledge that seems to be called for is knowledge of the causes of these facts. So to know why the kettle is boiling is to know that it has been heated, to know why the door will not open is to know that it is locked and so on. These extra bits of knowledge are also knowledge *that*, but they are in a special relation to our original knowledge. They are knowledge of the causes of the things we originally knew. But when we look a little deeper into Aristotle's ideas we find that it is not quite cause in our modern sense that he expects in answer to the question " Why ? ". Sometimes it is rather that he expects to be told the nature of the thing or material in question. If we ask why sodium reacts with water our answer is not the cause of its doing so, but rather runs something like this : " Sodium is an alkali metal, occupying a certain position in the electrochemical series, and so it reacts with water ". The question " Why ? " is answered by our being told what sort of stuff sodium is. If we know the "essence" of sodium we will be able to tell how it will react, and be able to explain how it does react.

There are other kinds of answers to " Why ? " questions too. Aristotle gives this classification of possible answers :

1. The nature or essence of the thing or material involved.

2. The material of which a thing is made. This may sound a little strange to our ears, but think about this sort of case : " Why does the small grey ball weigh more than the large brown cube ? " "Because it is made of lead."

3. The source of change, and this is what *we* call a cause of a happening. The kettle changes from cool to boiling and the cause of its doing so is being on the fire, and that is the source of the change.

4. Aristotle considers yet another sort of answer to a " Why ? " question. In this we are asking for the purpose or reason of something. It is a question particularly appropriate to what people do. " Why did you put a coin in the slot ? " " Because I wanted some chocolate." This idea can be extended. " Why do frogs have webbed feet ? " " So they can swim more easily." Here this way of asking and answering questions pushes us onto dangerous ground. In the chocolate case my answer refers to a definite plan I had formed before you asked me the question, a plan which involves putting in a coin to get something. In other words my act of putting in the coin is purposive. There is nothing, so far as we know, purposive about a frog coming to have webbed feet, and it would give a very misleading idea of evolution if we talked in this way. Since a frog is an integrated organism we can ask for the function of each part of the animal, but we must be careful not to slip from that to thinking of the development of the frog as something purposive.

A scientist doing a study of something along Aristotle's lines deliberately sought answers to each of these kinds of " Why ? " question. When he thought he had found answers to each kind he thought that he had finished the study he was working on, because, so he supposed, there was nothing else to find out.

Finding the right answers to such questions is not such an easy thing as it looks, and in the *Posterior Analytics* Aristotle offers further guidance on the kind of questions to put to nature. Suppose we ask whether Mars has a magnetic field. The first step is to try to find whether or not it does, that is to find out whether our conjecture is a fact. Having done that we can then ask one of our " Why ? " questions of it, because we will then want to know the cause of that planet's having a magnetic field. To answer this we may have to find out something about the material of the planet. It might be discovered, for example that the planet was largely composed of iron. Or we may go further and ask what it is in the nature of iron that leads to iron things having magnetic fields. Nowadays we would set about answering the last question by investigating the structure of the atoms of the element, and try to find answers to our question in the way the electrons are arranged and how they behave. Knowing that sort of fact we would know why it was that anything iron could have a magnetic field, and hence why an iron planet would have. In a moment we shall see how this idea of method was fitted in with Aristotle's logical theories. But he had another kind of question which might be asked. I might ask whether something which has been mentioned exists, and this might be a particular thing, say a particular man, or it might be some kind of material, or it might even be some process or even property. Thus we might ask whether Prester John exists, whether the Elixir of Life exists, whether there is a way of transmuting lead into gold, whether there is any substance which has the power to cure all diseases. If we do decide that whatever we have thought of does exist, then the next question is to decide what sort of thing it is, particularly what sort of nature it has, because we can then form some sort of idea as to what it will do, that is how it will behave in different circumstances. Or if it is a process or a property, what sort of effects it will be connected with. So we can see that the final step in both the kinds of investigation Aristotle mentions is to try to find out the nature of the things to which our answers to the preliminary questions direct us. This corresponds to our own efforts these days to find out chemical constitutions of substances, to find out the atomic structure of elements, to find out the crystalline structure of metals, and such things. It is Aristotle's view that " the nature of the thing and the reasons for the fact are identical ". So the reasons why things have the properties they do, and behave the way they do, are to be found in a study of the natures of these things.

How is this knowledge, once found, to be organized ? This brings us to Aristotle's ideas about logic. Logic is the theory of the organization of knowledge. To understand Aristotle's view of the nature of scientific knowledge and its organization we must first learn a little about his general theory of logic, in particular what his technical terms mean. The basic structure of demonstrative reasoning is, in his theory, the *syllogism*. This is a particular arrangement of statements into which more complex reasoning will be analysed. A single syllogism is made up of three statements, two of which are premises and the third the conclusion which follows from the premises if the syllogism is valid. A syllogism might go like this :

No trains are now coal fired
Only coal fired trains are dirty
therefore
No trains are now dirty.

You will see that three items are mentioned in the argument. The trains are the first item mentioned, then those which are coal fired, and finally those which are dirty. The syllogism tells us how these are related. You will notice that though there are three terms or words mentioning specific items, in the premises, that is the first two statements, there are only two terms in the conclusion. One of the terms then is functioning as part of the reasoning, not appearing in the conclusion. We could represent the structure of our syllogism in a sort of formula, which would make this clearer still. We could say :

No *T*s are *C*s
All Ds are Cs
No Ts are Ds

T and *D* appear both in the premises and in the conclusion, but *C* appears only in the premises. *C* is called the *middle term.* It is what binds the two premises together, and allows the conclusion to be drawn. It is what establishes a connection between *T* and *D*, the terms of the conclusion.

Aristotle thinks that in science we are working our way back up through syllogisms from facts to their reasons, from conclusions to premises. The facts we know are the conclusions we wish to prove, so the job of a scientist is to find suitable premises from which to construct suitable syllogisms. Suppose I know that hydrogen and oxygen combine in the ratio 2 : 1 by volume. I discover this by experiment. I might strive to increase my scientific knowledge by trying to find out why. The result of my discoveries and conjectures could be expressed this way :

At the same temperature and pressure the same volumes of all gases contain the same number of molecules.
Two molecules of hydrogen combine with one molecule of oxygen.
Hydrogen and oxygen combine in the ratio 2 : 1 by volume.

We have built up a syllogism from which the fact which we already knew follows. Nowadays we would say that we have found a theoretical explanation for the fact. What did we have to add to do this ? We knew how to recognize hydrogen and oxygen as distinct gases, and to measure the ratio of volumes which combine. We have added the idea of molecules. In Aristotle's language we have added a middle term. Trying to build a theory then is searching for a middle term. He connects up what he has to say about the kinds of question which should be asked by scientists, with the search for the middle term, by pointing out that the answer to the question about the reason of a fact and the question about the nature of the things and processes involved gives one the middle term. In our example the two premises give the reasons why the combining ratio of volumes is 2: 1, at a low level of explanation, and introduce a molecule theory of the nature of gases.

Not all theories are as simple as the one we have just been looking at, and they usually involve stacking syllogism on syllogism and pursuing middle term after middle term. Let us look at some further steps in our example. Suppose we want to know why the *molecules* of hydrogen and oxygen react in the ratio assumed in the theory, and we find an explanation through the middle terms " valency " and " atom " in the following syllogism :

One divalent atom usually combines with two monovalent atoms
The atoms in hydrogen molecules are monovalent, and those in oxygen molecules divalent.
Two molecules of hydrogen combine with one molecule of oxygen.

We may now want to go further and ask why is it that having such and such a valency makes it possible for molecules to combine in certain ratios only. We introduce another middle term " structure ". Now we get :

All atoms with a certain structure are monovalent.
Hydrogen atoms have such a structure
Hydrogen atoms are monovalent.
All atoms with a certain other structure are divalent.
Oxygen atoms have such a structure.
Oxygen atoms are divalent

(We may want to go further still and introduce electrons into the structures. There are many steps from this point up to present ideas—but we have started in the right direction.)

Such a stack of syllogisms is called a *sorites*, and it has in this case, several middle terms, " molecule ", " atom ", " valency ", " structure " and so on. One could imagine further middle terms being introduced and the sorites getting longer and longer. Surely, one supposes, there must be some end to this process. We do not want our theories to be endless stacks of propositions.

Aristotle did have an idea about how the search for middle terms could be brought to an end. To understand this idea of his we have to introduce another logical notion. It is that of a definition. A definition does not need to be proved, it is what is given or assumed. One cannot ask for a proof of a definition. So there is no point in looking for middle terms to build up a syllogism by which to provide such a proof. If we could find a sorites which went back to a definition that would bring an end to the process of finding middle terms and building syllogisms. There would be no further back to go. That is Aristotle's idea as to how to find an end to the process of building up an explanation. In the tidy way philosophers have he thinks that the self-same process of trying to find out the natures of the things involved in the causes of the facts to be explained will also give us the key to the definitions we need to halt the otherwise endless process of explanation. And this is for the very simple reason that he thinks when we say what the true nature of a substance, a thing, a process or a property is, we have defined it. The definition of hydrogen would then be that it is an element whose atoms are composed of a single proton and a single electron. From this defintion we should be able to fill out, using middle terms connecting having that structure with various powers and properties that an element might have, all facts that are true of hydrogen, such as for example that two molecules of hydrogen combine with one of oxygen, that the molecules of hydrogen in normal circumstances are diatomic, and so on.

You can see that this is really not too bad an account of science, though in some ways it is a bit too simple. There is nothing in Aristotle's works on the subject to tell us *how* to find middle terms, and how to judge when we have found the right ones. What if we were to construct a syllogism like this :

Paper is made of lead
Lead ignites at 240°C
Paper ignites at 240°C

Given the premises, the conclusion certainly follows, and the conclusion is certainly true. But this would be ridiculous as a piece of science, partly because we know that the premises are both false. So it cannot be just a matter of finding middle terms and constructing syllogisms. What the syllogisms say must be sound. We shall see that it is not so easy to tell whether a theory is a good one or not, though we can usually tell if a theory is bad, when it gives the wrong results. But that syllogism shows that some bad theories give the right results. So the fact that a theory gives the wrong results is not the only reason for decrying it. Sometimes too when a theory gives the wrong results to start with we keep on with it, not wanting to discard it, trying instead to improve it and to make it work properly because we feel there is something right about it. Later on, in examining Gilbert's theories, we shall have occasion to look into the reasons which prompt people to find one kind of theory acceptable and another not.

However incomplete Aristotle's account may be it is not a foolish one, and I now want to look at a fine piece of research done in the Middle Ages by a man called Theodoric of Freiberg. This will illustrate Aristotle's ideas being followed in practice and help too, I hope, to get rid of the all too common idea that science began with Galileo.

Theodoric was probably born before 1250 and died some time after 1310. He had a successful career in the Church, though he did not rise to particularly high office. In the year 1304 the Master General of the Dominican Chapter, Aymeric, asked Theodoric to write a work on the causes of the rainbow and other radiant phenomena. This Theodoric did under the title *De Iride et radialibus impressionibus*, i.e. *Concerning the Rainbow and Radiant Phenomena*. In this work Theodoric not only offers a theory to account for the position, shape and so on of the rainbow in terms of the refraction of light in raindrops, but he also offers a theory to account for the production of colours in the course of the process of refraction. It is to this theory that I shall first turn.

Theodoric's theory of colours is a curious one to our ears. He does not have the idea that white light is made up of coloured lights, but that colour is imposed upon white light by some physical process through which the light goes. Light acquires colour in some process like cloth acquires colour by being dyed. He devises a theory by which this process of acquiring colour can be understood. The style of the theory reflects another very important idea about nature and natural processes of great importance to Aristotle and his followers. This is the idea of contrary principles. Aristotle seems to have believed that there were four principles, the hot principle, the cold principle, the dry principle and the wet principle, and that the properties of things were determined by the proportions of their combination. Following this model Theodoric looked for contraries in the processes by which he supposed colours to be generated. He noticed that " radiant colours ", that is colours produced by refraction and by passage through a coloured medium, are produced in bodies which are more or less translucent, and which range from the clear translucency of water and glass to the obscure translucency of nacreous substances like alabaster. Translucency corresponds to the formal cause of colour, and he looks for another pair of contraries to provide the material cause of colour. Things differ not only in how translucent they are but also in whether or not they let light through their surfaces into their interior. A mirror is a perfectly bounded thing, because it lets no light through its surface, while a piece of glass is perfectly unbounded, since it lets all of the light through in some circumstances. But Theodoric knew about internal reflection and he makes use of the idea in explaining the rainbow. So he has to say of drops of water and spheres of glass that they are neither perfectly bounded, nor are they perfectly unbounded. Light can get into a raindrop, and is refracted in making entry, but then is internally reflected at least once, and at that point the raindrop must be supposed bounded. So we have two pairs of contraries, raindrops are more or less translucent, and more or less bounded like most things. When rays of light are refracted the perfect balance between the four contraries is upset, and the rays appear as coloured light. Whiteness is the result of a perfect balance, not between colours as the modern theory has it, but in the balance that Theodoric thought to obtain between principles. It is the imbalance between the principles that generates the colours when they meet someone's eye, or fall on an opaque surface such as a screen or a wall. The fact to be explained is the appearance of the colours, on refraction. The middle term is the idea of the two contrary principles, and this is an explanation because it leads back to a definition of colour, its essence or true nature, which he takes from Aristotle, that colour is the extremity of the translucent in a bounded body.

In these parts of the study Theodoric seems to be thinking in a very unfamiliar frame of reference. But in the next stage of his investigation another and to our eyes more modern side of medieval science begins to appear, namely experiment. The difficulties of experimenting with water drops themselves, and perhaps the idea that the results of his theory should be checked in a wide range of cases, since the theory is supposed to offer a general view of colour generation, led Theodoric to devise a series of experiments with models of raindrops, that is with

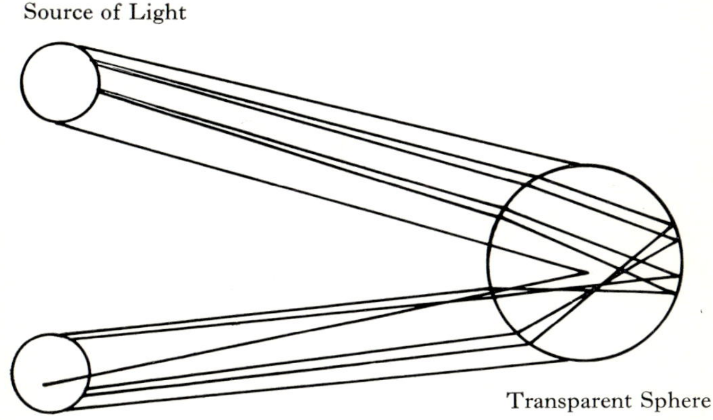

Fig. 1. After *De Iride*, by Theodoric of Freiberg.

large transparent spheres, either flasks filled with water or spheres of glass. Using a series of flasks he shows that there are two " modes of radiation ", that is two ways in which the refraction and generation of colours occurs in the spheres. If the ray enters fairly low down the circumference the light is refracted and then refracted again and the colours are seen with a certain order, such that if the sphere is to the left of the experimenter, red is seen farthest to the left. But if the ray enters fairly high up the circumference there is internal reflection within the sphere, and the colours are reversed, so that blue is seen farthest from the observer. These results are taken by Theodoric as " signs " of the correctness of his theory, that is as observations which tend to confirm the theory, since they are the effects to be expected if the theory is correct (fig. 1).

According to his theory the colours are created in different parts of the spheres and raindrops, red being a less clear colour, being formed nearer the bounded surface, while blue being a clearer colour is formed nearer the centre of the sphere, where transparency is greatest. The more bounded the place where the colour is formed the redder it will be, and the more transparent the place the bluer. The experiments with the flasks are seen by him as signs of the correctness of this theory.

The rainbow is generated in raindrops like the radiant colours he has been studying are generated in his glass sphere model. So the same effects should be found with each raindrop, the totality of which accounts for the way the rainbow appears. So following the schema of Aristotle for a scientific explanation we look for :

1. The Formal Cause : the form is a bow of radiant colours in the atmosphere.
2. The Material Cause : a cloud of drops of a spherical shape either stationary or falling. It makes no difference which, since in a cloud of falling drops one will take the place of another quickly enough for us to be justified in treating the phenomenon as if it had its causes in a cloud of stationary drops.
3. The Efficient Cause : this is the fact that a bright source of light (the Sun) is in the proper position to irradiate the drops, within which the light is affected by the imposition of the appropriate forms.

Light simply reflected from the outside surface of a drop will not be coloured, according to Theodoric's theory, so we have to consider only that light which is admitted to the interior of the drop and according to its passage through more or less bounded and more or less transparent parts of the medium acquires the form of this or that colour. The mathematical explanation of the position of the rainbow in the heavens with respect to the observer must then be based upon the consideration of rays of light which are internally reflected within the drop once or more often. Those which are reflected only once will form the primary bow, while those reflected round again will form the secondary bow and will have their colours reversed. Thus using the information derived from his models he forms a theory of the rainbow, and proceeds to geometrical construction to determine the angle of elevation of the high point of the bow above the horizon and the angular width of the bow. By a curious quirk Theodoric gives half values for each of these measures, namely 11° instead of 22° for the elevation of the bow, and 22° for 44°, the angular width of the bow. In a very similar way he studies the secondary bow, depending again upon his models to provide the basic verification of his way of thinking.

In finding material and formal causes of the phenomena Theodoric has answered those questions which Aristotle framed as the way of finding out middle terms for the construction of the sorites which " demonstrates " the phenomena.

Suppose the phenomenon is the fact that the observer sees the rainbow with the highest colour red, then we proceed as follows :

The rainbow is a radiant phenomenon caused by light passing through drops.

When light passes through drops colour is imposed upon it

The colour is imposed in such a way that the order of colours is red, yellow, green and blue.

The order is such that the colour reaching the observer from the highest drops is red, from the lowest blue.

With this order of colours the observer will see the highest colour as red.

> The rainbow is a radiant phenomenon which is seen so that its highest colour is red.

In this sorites we have demonstrated the fact by using a series of middle terms which enable us to connect the fact with the definition of the rainbow from which we began. This is a complete piece of science according to the Aristotelian point of view, and so it was regarded by those who accepted that ideal of scientific theory.

We shall see that little change of *method* is to be found in the science of the seventeenth century, but we will find that it is particular ideas like the Aristotelian theory of forms, and of contraries, and of the Four Principles, that are repudiated by the new generation of scientists from which our modern profession springs.

Further reading

An excellent account of Theodoric's work can be found in William A. Wallace, *The Scientific Methodology of Theodoric of Freiberg*, Fribourg University Press. A brief account of this work can be found in A. C. Crombie, *Medieval and Early Modern Science* (Anchor A167 a and b), pp. 110–2. For other medieval science see M. Clagett, *Science of Mechanics in the Middle Ages* (Oxford), and for a brief account see Chapter Three of *The Sciences, their Origin and Methods*, edited by R. Harré (Blackies). For an idea of the Greek view of the world and of their science one can consult B. Faringdon, *Greek Science* (Penguin), or S. Sambursky, *The Physical World of the Greeks*.

Theodoric's experiments

1. For one who believed that colour was produced by the effect of Four Contraries, there should be four main colours, one from the part of a medium that was less transparent and more bounded, red ; one from the more transparent but more bounded part, yellow ; another from where the medium was less transparent and less bounded, green ; and finally blue, which would be a form imposed upon light where the medium was less bounded but more transparent. Traditionally it had been supposed that there were only three colours blue, green and red. Repeat Theodoric's experiments on this point. In one he looked at a drop of dew on a spider's web in the early morning, with the sun behind him, and even as close as two finger breadths from the drop, he was able to see yellow as a distinct colour by moving his head with respect to the drop. In another he looked at the spectrum formed by passing sunlight through a prism onto a white surface, where again he saw yellow as a distinct colour, along with red, green and blue. How

many distinct colours do you think there are ? Why do you think Theodoric thought there were only four ?

2. To show that drops of water are responsible for the rainbow, and not any property of the air, Theodoric stood under a tree, with the sun behind him, and flicked droplets of water out of a bowl into the sunlight. See if you can produce a rainbow in a similaı way. Perhaps you could use the fine spray from a hose.

3. As I have described, Theodoric used spherical flasks full of water to simulate raindrops. With a bright light behind you see if you can verify Theodoric's dicovery that if the flask is to your left, red will appear at the extreme left when the ray passes close to the centre of the sphere, but that when the ray is passing higher up through the sphere the colours will be reversed. Compare Descartes' mathematical explanation of the phenomenon in terms of minimum deviation with that of Theodoric's. Does it matter which explanation we adopt so long as we can explain the rainbow ?

Draw a diagram to represent the way you think the rays are passing through the flask. Compare your diagram with that of Theodoric, as reproduced in fig. 1.

CHAPTER 2

the magical tradition and atomism

In 1460 a monk arrived in Florence from Macedonia, bringing with him a copy of a collection of supposedly ancient books, written by a mysterious and wonderful person called Hermes Trismegistus. At once Cosimo de' Medici ordered work on the translation of Plato to be stopped and the Hermetic books to be translated instead. To understand the excitement caused by the discovery of the Hermetic books we must realize who and what the people of the fifteenth century thought their author was.

Hermes, it was believed, was an early Egyptian priest, living long before Plato and long before Christ. He was the master of three arts, that of religion, of politics and of magic. Hence his title, Trismegistus. He was the first and greatest theologian, law-giver and magician. In those days people believed that the state of knowledge had declined since the ancient days, and the most ancient books of all would contain the greatest knowledge and wisdom. It was partly because Hermes was supposed to be so exceedingly ancient that his books had acquired such a reputation. It was partly because he seemed clearly to anticipate the coming of Christ, and the theories of Plato. Such a man could not be evil. So the magic which was also associated with his name must be an acceptable kind of magic too. The discovery of the Hermetic books seemed to justify Christianity, Platonic philosophy and natural magic all at once. It is no wonder then that Cosimo wanted Plato set aside for this more ancient and more wonderful sage.

The idea that there was a natural and good magic and a black and evil magic is a very old one indeed, and was made in Europe long before Hermes' works appeared. Natural magic was not only magic used for good purposes, it was practical, and often amusing, and in its more exciting aspects relied upon the conjuring of good spirits. In Shakespeare's *Tempest* Prospero is represented as a good magician using the spirits of the air to get good things done, and subduing the dark and evil Caliban.

Magic was a way of getting things done. It was an art with intensely practical ambitions, whether these were good or evil. It was no good if it did not work. Gradually practical advice and recipes and instructions for doing things began to be incorporated amongst the formulae of the older magic of spells. By the middle of the sixteenth century recipes and instructions for practical technology had been separated out into treatises of " Natural Magic ", amongst which are descriptions of things one can do with the magnet, as well as directions for lighting stages for plays, recipes for medicines and so on. In the new books of natural magic Aristotelian explanations of the phenomena appeared, later to excite Gilbert's scorn. This strand of practical knowledge became interwoven with the origins of modern science, and we shall return to it later. There was another strand from the magical tradition of no less importance.

The Hermetic magicians had a powerfully developed theory as to how the Universe worked, and indeed how men fitted into the whole system of the world. They believed that all the things in the world were really alive, in that whenever anything happened it happened because of the operation of a spirit. In Book IX of the *Pimander* as translated by Everard in 1650 we find the following dialogue between Hermes and Asclepius.

Hermes The motion then of the Worlds, and of every material living thing, happeneth not to be done by those things that are without the world, but by those things within it, a Soul or Spirit, or some other unbodily thing, to those things which are without it. For an inanimated body doth not move, much less a body if it be wholly inanimate.

Asclepius What meaneth thou by this O Trismegistus ? Wood and stones and all other inanimate things, are they not moving bodies ?

Hermes By no means O Asclepius. For that within the body which moves the inanimate thing is not the body, that moves both, as well the body of that which beareth, as the body of that which is born ; for one dead or inanimate thing cannot move another ; that which moveth must needs be alive if it move (p. 109 of 1650 edition.)

We shall find this theme re-echoing strongly in Gilbert. And not only in him. Throughout the period from 1450 to 1750 or thereabouts there was a continuous controversy as to whether inanimate things can have the power to act of themselves, or must their powers be explained by looking for the interference of a spirit ? There is no doubt what the magicians thought.

Not only were the things in the world alive or at least moved only when animated by a spirit, but it was also believed that there were " influences " streaming from one thing to another, when the things were of a like *kind.* And here the magicians picked up a theme already present in the alchemical tradition. This connection is not surprising because in fact the works of Hermes Trismegistus and the alchemical tradition stemmed not from Ancient Egypt at all, but from the Egypt of the post Christian era, and both had their roots in a movement called neo-Platonism. Likenesses of kind were not obvious, but derived from a complex theory which collected things around the signs of the Zodiac and the planets. There were, for instance, Saturnian metals, plant's, gems, colours and so on. And there were also Saturnian images. By using these principles a magician was supposed to be able to draw down the influences of the planet of his choice into the images and the things of the appropriate kind. When Hermes' book turned up it was found, to the delight of all with this way of thinking, that this greatest and most ancient authority himself had constructed statues in Egypt which drew down the influences of the stars, and could even speak. This idea too became the common property of scientist and magician alike. We shall find that Gilbert's theory of magnetism though it is not a Hermetic theory uses the idea of likeness of kind. Nor is it too fanciful to suppose that Kepler's idea of Universal gravity as an influence between all material bodies, made so much of by Newton, is a descendant of the more occult and mysterious influences of the Hermetists. Things could act upon one another at a distance without the intermediary of anything material because they were things of a like kind.

The magicians had a further elaboration of this theme, which influenced Gilbert in his studies of the Earth's magnetism, but became more and more attenuated as science proper developed. It was the idea of the connection and parallelism between the macrocosm, the whole Universe, and the microcosm, Man. Man was supposed to reflect the structure of the whole universe. This idea had various, more or less sophisticated forms. Here is one of its more naive manifestations, as explained by Hermes Trismegistus in Book X, § 11 of the *Pimander*. I quote from Chambers' translation of 1882.

" For every material movement is generation. The mental state moves the material movement in this manner ; since the world is a sphere, that is a head ; but above the head there is nothing material ; just as neither is there anything mental beneath the feet, but all material. But Mind, head itself is moved spherically, that is similarly to a head. As many things then as are united to the membrane of this head wherein is the soul, are by nature immortal. . . . But things distant from the membrane, in which the body possesses more than the soul [does], are by nature mortal ; but all is a living animal ; so that the universe is composed both of material and mental."

Many people believed that it was possible by a course of study to direct and remake one's own mind so that it became more and more a microcosm and so more and more like the world. In this way the influences of the Universe would stream down into one. To do this one had to maintain an awareness of the kinds of images which I mentioned when talking about the principles of likeness of kind. This had applications in medicine and in attempts to gain knowledge. Ficino, the man who first translated the words of Hermes, was a doctor and warmly recommended the use of talismans, images, colours and so on for curing mental problems particularly. He discusses the best way of dealing with the kinds of depression which students get into about their studies. One should try to draw down the influences from the cheerful and happy planets, particularly Jupiter, Venus and the Sun. It would be best, he thinks, to have constructed a general image of the world in which the three cheerful planets are emphasized. It should be made of brass, gold and silver which are the proper metals of the three planets, and should use the colours green, gold and blue, for these are the appropriate colours. One might finish up with something enamelled, showing a blue sky with golden stars, and the figure of Ceres dressed in green. By contemplating this object the spiritual influences from the three good planets would be drawn down and concentrated in one's own self. Ficino also advises that bedroom ceilings should be painted with appropriate health and happiness producing images, partly so that when one goes out into the ordinary world one carries these images with one, and so concentrates upon and notices things particularly consonant with them.

Much the same idea lies behind the various attempts to make knowledge " machines ". A most elaborate project of this sort was undertaken by Guilo Camillo (born 1480). He had the idea of constructing a miniature theatre in which the appropriate images, talismans, colours, gems, metals, etc. were arranged with drawers underneath them, into which inspired writings would be put. The seeker after knowledge sat within the semi-circle of tiers of images and so focused their influences upon himself. Whatever he wrote down while in this position was knowledge drawn from the macrocosmos itself by the images.

Other seekers after knowledge dispensed with the actual images, drawing instead upon the influence-concentrating powers of mental images. Giordano Bruno elaborated an immensely complicated scheme for ordering and controlling appropriate images and so of acquiring knowledge of the macrocosm. In general justification of this procedure many people returned to a famous passage in the writings of that supposedly ancient sage Hermes Trismegistus. There he speaks of images which spoke and which were animated by spirits drawn down from the macrocosmos. Other authors in the early Middle Ages attributed similar wonders to him. Now one must remember that many people in Renaissance times thought that Hermes was an important prophet of Christianity, and indeed used the apparent antiquity of Hermes' remarks which foreshadow Christ as a proof of the truth of Christianity and its claims. So when such a religiously authenticated man was supposed to have performed magic operations of the influence-drawing kind this could be taken as a justification for doing the same thing themselves, without laying themselves open to any accusation of trafficking with the devil. Hermetic magic was good magic because Hermes was an authentic Christian prophet. In this rather tangled way the general idea of influences streaming through the Universe between things of a like kind became fairly fixed in the European set of ideas, to be transformed slowly into such concepts as magnetic attraction and Universal gravity. In a sense Newton was the last great magician, as well as the first great scientist, since his system was based upon the idea of Universal gravity. We shall come to see, when we study Gilbert's ideas about magnetism, how profoundly the idea of influences acting among things of like kind determined his theory of magnetic coition.

We can sum up this strand of the magical tradition with three propositions from the works of Pico della Mirandola, a pupil and younger contemporary of Ficino. Amongst his *Conclusiones Magicae* we came upon the following :

2. Natural magic is allowed and not prohibited.

3. Magic is the practical part of natural science.

9. Only natural magic and Cabala are magics which testify to the Divinity of Christ.

13. Doing magic is nothing else but making a marriage with the stars.

Respectability had come to magic with the appearance of the Hermetic writings, and natural magic gradually began to come under more and more rigorous pragmatic demands as the practice of magic became more open if not more widespread. Magic had to be made to work. In this way the greater respectability of magic led to an emphasis on that part called " natural magic " in which natural forces were used to achieve practical effects. At first, as we saw with Ficino, it was uncertain just what were the natural forces upon which one could draw. Many natural magicians thought that the influences from the stars and planets were natural forces, and that there was no difference in principle between manipulating them for practical ends and, for example, using the power of water in a water-wheel. The magnet was often used as an example of something possessing a power whose mode of operation was as invisible as the influences from the stars, but which had as definite and practical an effect in the movement of things as one could wish. Gradually these influences were sorted out, those which were genuine forces and influences being slowly distinguished from those which were not. By the middle of the sixteenth century something very close to our own distinctions had been hammered out. Even so there were many who believed that one of the main tenets of astrology, the alleged influence of the situation of the stars and planets on people's character, was a genuine influence. Kepler, who did much to fix our modern ideas of what were genuine influences and what were not, was himself not entirely sure about the characteristic astrological influences even as late as 1600. He believed that light, magnetism and Universal Gravity (a concept he himself introduced into physics) were genuine influences propagated from one place in the Universe to another. He was sure that particular happenings in people's lives were not determined by the stars, but he remained uncertain about possible general effects upon human life which might emanate from the stars. We will hear in the next section of the rival theory of atomism, and how in the seventeenth century the two main views of nature were welded into an uneasy synthesis.

To give you some idea of the scope of sixteenth century " natural magic " I shall describe in some detail the famous book on that topic by Battista della Porta, because in such books Gilbert found some of the early theories of magnetism which we shall follow him in rejecting.

We can also get some idea of the difference in the style and ambitions of magic between his time and that of Ficino and Pico della Mirandola. Just like Pico, Porta begins by distinguishing between his magic and " a magic which is infamous, and unhappy ; because it hath to do with foul spirits, and consists of Inchantments and wicked curiosity ; and this is . . . an art which all learned and good men detest ". For him magic is the " practical part of natural philosophy . . . and a magician is such a one as works by the help of nature only ". Already we can see the emphasis on the practical, the distinction from black magic, and the confinement of method to using nature. If not by sympathetic influences, how does Porta see that things do happen ? Not surprisingly he offers an Aristotelian explanation. Aristotle uses the ideas of a natural balance between the Four Elements, and the distinction between Matter and Form in his account of how his questions should be answered. " Every natural substance ", says Porta, " is a compound of matter and form, as of her principles . . . [there are] three principles . . . unless some were stronger than others their virtues could not be perceived. . . . Neither yet is the matter quite destitute of all force . . . the effects of the qualities come from the temperature of the Elements . . . but the effects of the matter from the consistence or substance of them."

By " temperature of the Elements " Porta means their balance or equilibrium, so that things differ as to whether hot or cold, wet or dry predominates in their constitution. But he also believes that the forms are in some sense substantial, that the forms have their own kind of being and do not depend upon being the forms of particular things to exist. The form is what is fundamental. It is not easy for us to conceive of the Renaissance conception of " Substantial Form ". Here is what Porta himself says about it. " But the Form hath such singular virtue that whatsoever effects we see all of them first proceed from thence and it hath a divine beginning . . . being the deepest part she useth the rest as her instruments . . . the temperature and the matter . . . are but as it were the instruments by which the form worketh." These mysterious forms, though they come originally from Plato's ideas, have been subtly transmuted into the system of the magicians. Porta says : " The form comes from heaven received from the intelligences. . . . The soul of the world or Form-giver passes the Form to the Stars and intelligences and so down to the Four Elements." But having been so transmuted they serve, in Porta's view as an explanation of Aristotle's system of the world. We shall return to an important application of this point of view, described by Gilbert as " the maunderings of a babbling hag, rather than the devices of an accomplished prestigiator ", when we look at Gilbert's refutation of the theories of magnetic attraction of his predecessors. It is enough to see now that by the time of Porta some sort of synthesis of Platonic, magical and Aristotelian ideas had been worked out. This synthesis was found as objectionable by the admirers of Hermes as it was by those who wished to revive the ideas of the atomists. Both parties fell upon the doctrine of substantial forms with derision. " As for the causes of magnetic movements, referred to in the schools of philosophers to the four elements and the prime qualities, these we leave for roaches and moths to prey upon ", says Gilbert.

Whatever may be the justice of these remarks there is no doubt that a substantial body of practical technique and some information of a more general kind were incorporated in the treatises of natural magic. The importance of the demand that its recipes and instructions *worked* can hardly be over-emphasized because it forced its practitioners to pay attention to the way things actually behaved. It was thus an important element in the rise of experimental science. We shall look at some of the techniques described in Porta's famous book. The examples are taken from an English edition, published in London in 1658. There are many paragraphs devoted to practical horticultural advice. For example, he explains how gourds and cucumbers can be made to bear seedless fruit. " The Gourd ", say they, " will grow seedless, if you take the first branch or sprig of a Gourd when it is a little grown up, and bury it in the Earth as they use to deal with Vines, so that only the head thereof may appear ; and so soon as it is grown up to bury it so again : but we must take special care that the slips which grow up out of the stalk be cut away, and none but the stalk be left behind ; so shall the fruit that grows upon it, whether it be Gourds or cucumbers be destitute of all seed within. Likewise they will grow without seeds in them, if the seeds which are planted, be macerated or steeped in Sea Salmon Oil, for the space of three days before they be sowed."

This is a far cry from the muttering of spells, or the placement of green and gold images in the bedroom of a depressed student. Some of Porta's recipes are for no more elevated purpose than practical jokes. He describes a way of making brass seem to be silver by rubbing it between your hands. " As boyes or cozening companions ", he says on p. 168, " are oftentimes wont to do, that if they do but handle any vessel of brass they will make them straightways to glitter like silver, you may use this devise. Take ammoniac-salt, and alum, and salt-peter, of each of them an equal weight, and mingle them together, and put into them a small quantity of silver dust that hath been filed off ; then set them all to the fire, that they may be thoroughly hot ; and when the fume or vapour is exhaled from them that they have left reaking, make a powder of them, and whatsoever brass you cast that powder upon, if you do withal, either wet it with your own spittle, or else by little and little rub it over with your fingers, you will find that they [the metals] will seem to be of a silver colour."

There are included books on cooling, on refining metals, on medicine, and on cosmetics and perfumes. There are several pages of hair dyes for women, and he notes that women " think their hair the more beautiful, the more yellow, shiny and radiant it is ". There is even a chapter on Invisible Writing. One useful recipe for spies might be this :

" Write on an egg with juice of lemons or onions or fig-milk. When you put this to the fire the letters will appear yellow ; and that must be done with a raw egg, for if you boil it the letters will be seen ". Along with the strong element of practicality goes a strong element of scepticism. Speaking of the recipes of Albertus Magnus, Porta says that many of them do not work, because he has tried them. Many times he cites his own experience in support of a formula or practice. In the transition from Ficino to Porta we can see one of the lines of development that led to the experimental method.

Further reading

Detailed and interesting account of Hermetic magic and its influence can be found in Miss Frances Yates' two books, *Giordano Bruno and the Hermetic Tradition* and *The Art of Memory*. The ideas of the alchemists can be found in Charles Singer, *A Short History of Scientific Ideas*, pp. 101–2, 144–7, 184–6 ; and in A. C. Crombie, *Medieval and Early Modern Science*, Vol. I, p. 129–39. An interesting account of the microcosm/macrocosm idea, as well as of the use of magic in medicine can be found in Charles Singer, *From magic to science*.

Practical work

1. One of the last surviving traces of the magical tradition in our times is astrology. There are many predictions of an astrological kind in the appropriate columns of some newspapers. Take the predictions for one week for the members of your class, using your knowledge of their birthdays to find out under which zodiacal sign they fall ; determine the proportion of predictions which turn out to be correct. Can you spot anything in the way the predictions are made that makes it difficult to say whether or not they have been fulfilled ?

2. Many people believe that warts can be charmed away. Collect up as many magic spells and recipes for doing this as you can. Do they work ?

The theory of matter and forms was not the only world picture that the Renaissance inherited from antiquity. There was also the very influential ideas of atomism, and loosely connected with them the revival of mathematical techniques. I say " loosely connected " because the magicians also had their mathematics, much of which we now see to be bogus. But we must remember that at that time the natural philosophers were no more certain as to what was proper mathematics and what bogus than they were as to what were real influences in the Universe and what were not. Science arose partly, at least, because these differences were more or less successfully made clear.

But atomism was a general theory of the constitution of things and explanation of natural change that was a rival to the Aristotelian theory in generality and power. It arrived in Western Europe largely through the poem of Lucretius, the *De Rerum Natura*, written in the first century before Christ. The atomistic ideas around which the argument is built come from earlier Greek sources, notably the philosophy of Democritos, of the fourth century before Christ. The first arrival of atomism was unpropitious. At least, in the form given it by Lucretius, it is associated with doctrines of the soul that are hardly less than atheistic. Lucretius says explicitly at line 418 in W. E. Leonard's translation :" . . . minds and the light souls of all that live have mortal birth and death ". The doctrine of atoms was associated too with Epicurus, who was supposed to have advocated a life devoted to sensuous pleasure. It was because of these popular associations of atomism with atheism, in part, that the first efforts to introduce atomism into England failed. This failure is epitomized in the life of the great mathematician Thomas Hariot (1560–1621). With many others he was arrested in 1605 as a possible suspect in the managing of the plot of Guy Fawkes, which had had its unsuccessful culmination in November of that year. In 1592 Hariot had been denounced as an atheist by one Robert Parsons and the evil reputation stayed with him. To be known as an atomist was to be politically and religiously suspect. Atomism was reintroduced into England later in the seventeenth century through the ideas of the Corpuscularians, of whom Robert Boyle and John Locke were the pious and influential advocates. Its assimilation was greatly assisted by the comprehensive defence of both atomism and the Epicurian philosophy by the famous Abbé Gassendi. But all this was much later than the time of Gilbert and his studies of the magnet. Gilbert worked during the first period of atomism in England. He was not a comprehensive atomist, but did make great use of one of the ideas derived from that point of view, the idea of the effluvium. An effluvium was a swarm of minute particles, a kind of fluid mist, that could be exuded or squeezed out of material things, whence it would act upon other things in a way like a stream of particles, or like a wind, or sometimes as if the thing acted upon had been lapped around by a liquid. In Gilbert's time it was possible to have a particle view of one sort of phenomenon, and perhaps a Hermetic or even an Aristotelian view of the explanation of some different phenomenon, without feeling any particular strain. It was very different 100 years later. By then the corpuscularian view had become so deeply rooted that every kind of action and interaction was to be explained by some kind of material connection, a streaming and bombarding of particles. Newton, who had retained an essentially Hermetic idea in his theory of gravity as an action at a distance, was in most things an atomist, and even sought an atomic explanation of the action of gravity. In Query 19 appended to his *Opticks*, Newton mentions the idea of an aetherial medium penetrating transparent bodies and being involved in the passage of light. In Query 21 he asks : " Is not this Medium much rarer within the dense Bodies of the Sun, Stars, Planets and Comets, than in the empty celestial Spaces between them ? And in passing from them to great distances, doth it not grow denser and denser perpetually, and thereby cause the gravity of those great Bodies towards one another, and of their parts towards the Bodies ? "

The essential idea of atomism was that every material thing was made up of indivisible units or atoms, whose shape, motion and arrangement gave the body its various powers to act upon our sense organs and to act upon other bodies, particularly by contact. This picture required too that there be a void, a space quite empty of atoms within which the atoms could move and move freely until they met another atom. This idea ran strongly counter to certain views of Aristotle and indeed of the tradition, that there was no void, that the world was a plenum, full of something even where it seemed to be most empty. Most people in the seventeenth century came to deny this doctrine of the plenum, but despite the experiments of Torricelli, which had seemed to prove that there could be a vacuum, there were still some people, even one so eminent as Descartes, who clung to this idea. But at the end of the sixteenth century no clear body of doctrine nor any convinced group of upholders of any of the new ideas could be found. Gilbert, as we shall see, was very much a man of his time in calling upon a variety of ideas, of sources, as the need for them arose.

CHAPTER 3

the early history of magnetism

WILLIAM GILBERT was not only a great experimental scientist, but also a profound scholar of the origins of his subject. Some of his most beautiful experiments were performed to refute the theories of his predecessors in the study of the magnet, and the general direction which his more positive studies took was determined by earlier investigations. The three students of the magnet to whom he devotes the greatest attention were Peter Peregrinus, Battista della Porta and Robert Norman. In this chapter we shall examine the views of these men and find out what they discovered.

Permanently magnetized iron does not seem to have been known in antiquity. But there was wide knowledge of the lodestone, a rock or mineral, pieces of which have the power to attract iron. Our own word " magnet " seems to have come from the Greek expression *lithos magnetis*, referring to the name of a town, Magnesia, near Heraclea. The property of lodestones to attract iron was well known in antiquity and is mentioned by both Plato and Aristotle.

In his *De Anima* (405a) Aristotle cites Thales as one who believed " the lodestone to possess soul, because it attracts iron ". Plato mentions the power that the lodestone has of attracting iron, too, but he also notices that when a piece of iron is drawn to the magnetic stone it too acquires the power to attract iron, and so a whole chain of pieces of iron can be made to depend from a stone. The whole power manifested in each piece of iron he believes comes from the original stone. Plato, like everyone else who has considered magnetism, was puzzled by why the lodestone would draw the iron, why there seemed to be an " attraction " between stone and iron. In the *Timaeus* he does offer some sort of explanation, though it is not easy to see exactly what he thinks is happening. He says : " Moreover, as to the flowing of water, the fall of the thunderbolt, and the marvels that are observed about the attraction of amber and the Heraclean stones,—in none of these cases is there any attraction ; but he who investigates rightly, will find that such wonderful phenomena are attributable to the combination of certain conditions,—the non-existence of a vacuum, the fact that objects push one another round, and that they change places, passing severally into their proper positions as they are divided or combined " (*The Dialogues of Plato* : trans. by B. Jowett, Vol. III, p. 768). It looks as if Plato means that there is some kind of circulation of material, something which pushes upon whatever is being drawn to the magnetic stone, and that in some way the stone is responsible for the circulation. It seems clear that he is denying that there is any action at a distance directly between the stone and a piece of iron.

This idea seems to have been developed into rather more detail by other Greek philosopher-scientists. Empedocles is said to have had a theory which explained the apparent attraction as due to a physical stuff which passed from the lodestone to the iron, expelling the air from the pores of the lodestone, while an emanation from the iron passed to the lodestone. Then, thus joined, they were drawn together. A more detailed atomistic explanation was given by Democritos. He held that atoms were exchanged between iron and lodestone, because lodestone atoms expel iron atoms which enter the lodestone. In moving towards the lodestone the iron atoms cause the whole iron to move in the direction of the lodestone. Just why the atoms moved this way is not clear. It seems to be just as mysterious as the problem of why the whole piece of iron moves. Lucretius is however more specific. In Book VI of *De Rerum Natura* he says : " It will not be hard to give a clear account of why the magnet draws iron, and explain its cause given these principles. First a great number of seeds of lodestone stream off it and strike the air lying between the lodestone and the iron, and force it apart. When this space has been cleared and made a void then the primal atoms of iron fall into the vacuum, joined together, and because of this the ring of iron follows after them and moves as one piece. There is nothing more strongly coherent than strong iron. So it is no wonder that such a stuff follows those parts of it which fall off into the vacuum. The ring follows the atoms until it reaches the lodestone, and clings to it, with invisible links. This motion is assisted and made easier by another natural phenomenon. When the air is driven out from between the ring and lodestone and a vacuum formed between them then the air on the other side of the iron pushes it from behind. Air is always pushing on things which do not move, but in this case the iron moves because on the other side there is an empty space and the air can push the iron into it. The air gets into the pores of the iron and pushes it like the wind pushes a ship and sails. This happens in whatever direction there happens to be a void. Everything is moved by something else pushing upon it. And everything has some air in it because all things are to some degree porous, and air is around everything. So when the air in the iron is acted upon and becomes stormy it strikes the ring and shakes it. . . ." Then Lucretius gives some examples of the effect of the lodestone on iron, particularly the seething of a bowl full of iron filings when a lodestone is moved about underneath.

The Ancient World, it seems, was very well acquainted with the fact that lodestone draws iron, and even advanced theories as to how this happened, but they did not seem to have realized that lodestones can act upon each other, that iron can be magnetized, nor did they have the idea that the lodestone naturally tends to point north and south. These facts came to be understood later.

The fact that a freely suspended lodestone tends to remain in a north–south orientation was known to the Chinese and the Europeans in the twelfth century. The first mention of the phenomenon in the west is by Alexander Neckham, who describes a pivoted needle which takes up a regular orientation and can be used to determine direction in the darkness at night. Like many writers of the later Middle Ages and like the magicians of the early Renaissance, Neckham explained the magnetic phenomena in terms of similarities of kind between the attracting and attracted materials.

The first really systematic study of the magnet and the magnetic phenomena is the letter of Peter Peregrinus, written in 1269, though scholars do not think that he himself discovered the facts that he records so carefully. The letter is divided into two parts, one of which describes the magnetic phenomena, with a marked absence of any theory or explanation of them. In Part II some magnetic instruments are described.

Peter Peregrinus makes the polarity of the lodestone quite clear. He understands that there are two special points on the stone, somewhat like the north and south poles of the celestial sphere. He identifies one of these as a north pole, defined by him as the one which points towards the north pole of the celestial sphere, and the south pole is the one which points to the south pole of that sphere. That is a great sphere which contains the whole Universe, and at the centre of which lies the Earth. Upon it is the circle of the Zodiac, and the Pole Star marks its northern pole. It is to that pole that Peter Peregrinus thinks the freely suspended lodestone points.

These important polar points on the lodestone can be found by several methods. Peregrinus describes three ways of finding them. One is to lay little bits of wire upon a spherical lodestone and mark out lines which will intersect at the two poles. Another, is to find the points where iron is most strongly attracted, and the third is to see where a little short length of iron wire will stand upright.

Two further discoveries of great importance are found in this letter. There is a clear understanding that unlike poles attract one another, and that when like poles are presented to each other the magnet or the lodestone turns away. Peter, like other medieval and Renaissance writers, was unwilling to go so far as to say the like poles repel, and many people explained the apparent repulsion between like poles as really the attraction between the unlike. Peregrinus also knew some important generative relations between poles. If a lodestone is broken, then poles appear at either side of the break, so that a south pole will appear on the piece which has the original north pole in it, and a north pole on the piece which has the original south pole in it. The parts thus divided tend to come together again. In the magnetization of iron by a lodestone he was aware that the part of the needle touched with the south pole of the lodestone tends to point to the north celestial pole, and that the part touched with the north pole of the lodestone tends to point to the south celestial pole.

Peregrinus's three instruments include a lodestone mounted on a wooden float, in a graduated circular container, with a north point ; and fitted with an alidade for measuring the angular elevation of stars and planets. An alidade is a rod, pivoted at its centre, with vanes at each end. There are pin holes in the vanes, and the star is sighted through them. The second instrument is also a compass but with an iron needle fitted into a vertical pivoted axle, under a graduated glass or some other transparent cover. The compass needle is to be magnetized with the help of a lodestone. Peregrinus believed that a spherical lodestone mounted so as to be free to spin on an axis in the line of the axis of the celestial sphere would rotate and spin of its own accord. This seems to have been the basis of his idea for a perpetual motion machine. We shall come upon the same idea in Gilbert, where it is used to explain the daily rotation of the Earth.

Battista della Porta's account of the magnet is not nearly so modest, nor so clearly scientific. He explains the phenomenon by a theory of antipathy and sympathy and quotes both Aristotle and Anaxagoras with approval as believers in the living nature of the lodestone. His particular theory, the one Gilbert stigmatized as the " maunderings of a babbling hag ", invokes both love and strife. " But ", he says, " I think the lodestone is a mixture of stone and iron. . . . Yet do not think the stone is so changed into iron, as to lose its own nature, nor that the iron is so drowned in the stone, but it preserves itself, and whilst one labours to get the victory of the otheı, the attraction is made by the combat between them. In that body there is more of the stone than of iron, and therefore the iron, that it may not be subdued by the stone desires the force and company of the iron, that being not able to resist alone, it may be able by more help to defend itself. For all creatures defend their being . . . wherefore it willingly draws iron to it, or iron comes willingly to that." A slightly different point of view emerges in other sections of Book Seven of *Natural Magick*. In Chapter XX he says : " Because there is such a natural concord and sympathy between iron and the lodestone, as if they had made a league, that when the lode stone comes near the iron, the iron presently stirs, and runs to meet it, to be embraced by the lodestone ". And again in Chapter XXI he says : " The exceeding love of the iron with lodestone is greater and more effectual and far stronger than that of the lodestone with the lodestone ". The very same sympathy and antipathy idea is used by Porta to explain the fact, about which he is very clear, that like poles *repel* and unlike poles *attract*. He says that when unlike poles are close, one lodestone wants to " allure and attract it, and enjoy its company, and to foster it in its bosom, and again another should be such an enemy of it, that they are at mutual discord ".

Porta's discussion is really quite thorough, and he adds some points not found in Peter Peregrinus's letter. For instance, Porta makes some rough quantitative judgments about the force, and he thinks that the power of the lodestone is roughly related to its bulk. What is more, he gives a fairly detailed account of a method for measuring the force of a lodestone by a perfectly viable use of the balance, in which a lodestone is balanced against weights in the other pan, and then a piece of iron is put under the lodestone's pan, and then weights are added until balance is achieved again. The weight so added are a measure of the force of the lodestone. In general, on the subject of the force of the lodestone, he has some additional points to make. He notices that it can be passed to other lodestones, and that lodestones can form chains, much as pieces of iron can, dependent from a lodestone. The force is not affected by wood, or the material of a wall, or by brass or even by silver. But it is affected by iron, and a lodestone will not affect a needle lying on an iron plate if the stone is moved about underneath. He also made the important observation that there is a sphere of virtue around the lodestone, so that its virtue can be transferred to a piece of iron even when they are not in contact. It was this idea that Gilbert was to develop into the idea of the *orbis virtutis*, a clear forerunner of the nineteenth century idea of the magnetic field. Heating, Porta notes, destroys the virtue of the lodestone.

But perhaps the most striking feature of this work is the use Porta makes of experiments to refute the theories and false beliefs of his predecessors. He castigates them for standing in the way of learning by repeating uncritically the mistakes of the ancients. It had been believed and repeated by many authorities that garlic destroyed the power of the stone and of magnets generally. Porta showed that this was not so by rubbing a magnetized needle all over with oil of garlic and found that it made not the slightest difference to its magnetic power. It had also been told him that mariners did not eat garlic or onions for fear of upsetting the compass. Again he enquires. "When I enquired of mariners whether it were so, that they were forbid to eat onions and garlic for that reason ; they said, they were old wives' fables, and things ridiculous, and that seamen would sooner lose their lives, than abstain from eating onions and garlic." This scepticism he extended to the other great and ancient tale, that diamond would destroy the virtue of the lodestone. He tried it with tiny lodestones and big diamonds, and even with huge lodestones and tiny diamonds, and they had no effect whatsoever. Similarly he neatly brushes aside some of the views about the augmentation of the power of the stone. It had been said that by burying the stone in a heap of iron filings the stone would absorb the filings and become more powerful. He tried it and left the lodestone buried for some months, but when he measured its power it was practically the same. And when he tried boiling the lodestone with "oil of iron" he found that it lost its virtue altogether, as generally happened when the lodestone was heated.

On some other matters he was less sceptical. He believed, for instance, with many of his contemporaries, that the lodestone was specially concerned with love, and that it would " reconcile husband and wife ". But when a lodestone was burnt, or pieces of lodestone thrown on the fire they would give off fumes which would give one melancholy dreams, and even perhaps strike terror into people. He says that thieves sometimes put lodestone into domestic fires and then robbed the house when the inhabitants were frightened away.

Amongst the practical applications of the lodestone he of course mentions the mariner's compass, and describes the uses to which it is put, including its value in keeping the tunnels of miners straight, and assisting in the blowing up of fortifications and so on. But his joy is clearly in toys, and he describes two excellent magnetic toys. By breaking up a lodestone into different sized pieces one can form two armies, with foot soldiers and horsemen and others. Then these can be set out on a table and a most realistic battle can be made to take place by moving them about with lodestones held underneath the table. A rather more elaborate toy consists of a little wax boat, with a tiny wooden oarsman, with a magnet fixed to his feet, and little oars in his hands, pivoted on a hog's bristle. By moving a magnet under the vessel he can be made to move anywhere on the surface and rocks to and fro, seemingly rowing. By marking letters around the edge of the vessel he can be made to do fortune-telling by spelling out messages, and so Porta says, amusing women.

This is a careful, practical and charming account, and comes very close to real science.

In his deliberate use of experiment to refute " old wives' tales " Porta is surely on the very brink of systematic experimental science.

Robert Norman's contribution led directly to some of Gilbert's most important ideas. In making compasses Norman found that he had to add a counter-weight to the south end of the needle, or shorten the northern half of the needle, to overcome a tendency for the needle to dip or decline towards the north. If a needle, before magnetization is perfectly balanced on its pivot, after it has been touched by the lodestone and become magnetized it no longer balances in the horizontal plane but the northern end drops below the horizon. This phenomenon Norman called " declination ". He describes his discovery as being due to the anger which he felt on spoiling a needle by cutting off too much of the north half, and becoming determined to find out about the declination. Various questions occurred to him. How far would the needle dip or decline if free to move naturally in a vertical plane ? If there was a maximum angle, was this always the same at the same place ? Was the phenomenon due to an attraction between some terrestrial or celestial pole and the magnet's pole ? If not, what was the cause of declination ?

To answer these questions he performed two of the great classical experiments of the history of science. For the first he constructed a dip circle. He fitted up a graduated disc on a foot, so it could stand upright, with a plumb line to ensure that the disc was vertical. Then he fitted a needle with an axle, with sharpened ends, and rested these in two hollowed pieces of glass, one at the centre of his disc and the other supported by a piece of brass, also fixed to the foot. Thus he obtained a horizontally mounted needle free to move in a vertical plane. First the needle must be checked to see that it is free to move, and this is tested by swinging it and seeing that it stops anywhere round the circle. Then it is magnetized and declines out of the horizontal to set at 67° in London. Norman immediately conjectured that the constancy of the angle at London is no accident, and that there will also be constant dip at other places. What is more, he also conjectured that the dip will vary with the latitude in some way. Within a few years this prediction was triumphantly confirmed. But the puzzle of the cause of the declination remained.

The solution which Norman proposed depended really upon the introduction of a new concept, a new way of considering the whole phenomenon of the lodestone. On p. 13 of his little book, *The New Attractive*, London, 1581, Norman remarks as follows : " in my judgment [the point attractive] ought rather to be called the point respective ". In that remark field theory was born. Up until then the poles of the lodestone and compass needle had been thought to attract and repel each other, drawing the stone or the needle around with them. Instead of treating the poles as " points attractive " Norman proposes to treat them as " points respective ", that is to regard them as simply indicative of a directional phenomenon to which the whole magnet is subject. The magnet sets in a definite direction, and it is this *direction* which is characteristic of the various places on the Earth's surface. The reason why the magnet points north is not that there is a point on the Earth or in the Heavens that attracts a point on the magnet, but that the magnet tends to set in a north–south direction. The poles are points which *respect* the north–south direction, not points which are attracted to their respective poles. This is exactly the idea of the field. But despite the invention of the idea by Norman and its vigorous exposition by Gilbert the idea of points of attraction was later revived, particularly by Ampère. And Faraday had to overcome very considerable opposition from Ampère and others in his reintroduction of the field idea, expressed in his case in the concept of lines of force. But these are nothing but a concrete representation of the directional conception of magnetic phenomena clearly understood by both Norman and Gilbert. Norman's second classical experiment (fig. 2) was designed to show that the directional phenomena of the magnet were not attractive phenomena.

" Now to prove no attractive point neither beneath in the Earth, nor Heavens Northwards, nor above in the Heavens Southwards, you shall take a piece of Iron or Steel wire of two inches long or more, and thrust it into a piece of close cork, as big as you think may sufficiently bear the wire on the water, so as the same cork rest in the middle of the wire."

"Then you shall take a deep glass, bowl or cup, or other vessel, and fill it with fair water, setting it in some place, where it may rest quiet and out of the wind. This done cut the cork circumspectly by little and little, until the wire with the cork be so fitted, that it may remain under the superficies of the water two or three inches, both ends of the

Fig. 2. From the *De Magnete*. The classical experiment performed by Robert Norman and repeated by Gilbert to show that magnetism is a directional, and not an attractional phenomenon.

wire lying level with the superficies of the water, without ascending or descending, like to the beam of a pair of balances being equally poised at both ends. Then take out the same wire without moving the cork, and touch with the stone, the one end with the south of the Stone, and the other end with the north, and then set it again in the water, and you shall see it presently turn upon his own centre, showing the aforesaid declining property, without descending to the bottom, as by reason it should, if there were any attraction downwards, the lower part of the water being nearer the point than the superficies thereof."

So the idea that the north end of the needle is being pulled down by an attraction between the pole of the needle and some mysterious attractive point in the Earth or the Heavens is shown to be mistaken, since if that were so the needle would be pulled steadily down to the bottom of the glass. Instead we see a directional phenomenon, the whole needle setting in a north-south direction when viewed from above, and declining to 67° when seen from the side. It moves neither downwards nor across the bowl. Norman makes the final extension to the idea of the field in these words : " And surely, I am of opinion, that if this virtue could by any means be made visible to the eye of man, it would be found in a spherical form extending round the Stone in great compass, and the dead body of the Stone in the midst thereof, whose Centre is the Centre of his aforesaid virtue." So not only do we have the idea of a directional phenomenon in Norman, but also the idea that this virtue is distributed in space around the sources of magnetic power.

Further reading

An excellent summary of magnetic discoveries prior to the time of Gilbert is to be found in Duane Roller's book, *The De Magnete of William Gilbert*, Amsterdam, 1959, chapter 1. An excellent short account of the letter of Petrus Peregrinus can be found in A. C. Crombie's *Medieval and Early Modern Science*, Vol. 1, pp. 120–2. Indeed, in chapter 1 of the *De Magnete* itself, there is an admirable summary of the history of magnetism.

Practical work

Repeat Norman's experiment with the needle in the glass of water, if you are patient in paring away the cork. Find out whether we use the word " declination " nowadays for Norman's discovery.

CHAPTER 4

William Gilbert

I. *His life*

GILBERT was born at 2.20 p.m. on 24 May 1544. This is one of the very few facts known for certain about his life. His library and papers were destroyed in the Great Fire of London, as were all the records of the Royal College of Physicians, of which Gilbert was a prominent member. Gilbert's father was a lawyer and become Recorder of Colchester, in which town Gilbert was born. However, nothing is known about his early education. In 1558 he matriculated at Cambridge, and took his B.A. in 1560–1. He remained in Cambridge for some time, taking his M.D. in 1569, and in the same year he was elected to a senior fellowship. He remained at St. John's until some time after 1570, since we know he was appointed Senior Bursar in January of that year.

The next we hear of him, in the fragmentary records that remain, is as a member of the Royal College of Physicians in London in 1581. So at some time in those eleven years he gave up his fellowship and moved to London to practise medicine. From 1581 until his death in 1603 his career is fairly well known. He became quite prominent as a doctor, being amongst those mentioned as being capable of dealing with outbreaks of illness amongst the sailors of the English fleet, waiting in 1588 for the arrival of the Spanish Armada. He is very well spoken of in some letters of the time, and evidently had something of a fashionable practice, since he is known to have treated such persons as Lord Derby and Lady Cecil. He held various posts in the Royal College, becoming President in 1600. Also in that year he became one of the physicians appointed to the Court. One of his friends, John Chamberlain, complains in letters of the way in which this appointment took Gilbert away from his usual social circle, as well it might, considering the demanding nature of the Queen.

Gilbert never married and, according to one of his younger relatives, this was so that he could devote himself to the maintenance of his family. Unfortunately the portrait of himself which was given to the Schools Gallery in Oxford was destroyed. We do know however that he was a tall, self-contained and quietly cheerful man.

When, in all this medical business, could he have found time to do his magnetical experiments ? His book *De Magnete* was published in 1600 and as we will see contains the description of one of the most systematic and sustained pieces of experimental research ever undertaken by a single man. It seems evident that Gilbert must have done these experiments sometime between leaving Cambridge and coming to London, that is some time between 1570 and 1581. There does seem to be some evidence that Gilbert had friends among the Italian scientists, and it has been conjectured that he may have spent some time in Italy. In the present state of knowledge about Gilbert's life it is impossible to say whether he really did. But if we do not know much about Gilbert's biography we do know an extraordinary amount about his thoughts. The masterly interweaving of theory, experiment and interpretation of the *De Magnete* shows us the mind of one of the greatest of all scientists at work. We will follow his thoughts and in the practical work suggested we will do the experiments he did, and so perhaps come to get some idea of how science is actually done and of the status of its conclusions.

II. *The De Magnete, and the Nature of Magnetism*

The systematic experimental investigation of lodestones and their properties is subordinate throughout the book to the development of the consequences of Gilbert's great hypothesis, that the Earth itself is a lodestone, and his general theory of the fundamental character of magnetic coition, i.e., the power of magnetic bodies to draw together. His studies of electricity, interesting though they are, are really subordinate to the grand design in that they show how different, and how much less stable is electrical attraction from magnetic. We shall begin our study of the book by setting out Gilbert's ideas about the nature of magnetism, since it is against the background of these ideas that the whole of the work is to be understood. We will also bear in mind that Gilbert is the inheritor of the magical tradition, and that he imagines a world having some kind of unity with some fundamental mode of interaction amongst its parts. This fundamental mode of interaction is magnetic. For Gilbert iron is somehow connected with the most basic kind of stuff.

Like Robert Norman, Gilbert denies that there is magnetic attraction, that is one active body drawing another which is passive. This is expressed by him in his distinction between " coition " and " attraction ". Magnetic interaction is *co*-ition, a mutual action between bodies. It was the ignorance of the Ancients, so he thinks, that led them to think that attraction exists. This mistake he thinks leads to the bringing in of force and " a tyrannical violence rules ", that is one body acts upon another. Instead we must think of coition and " primary confluence ", a basic flowing together of things. What is this primary confluence ? We have already noticed in the first chapter the important distinction between matter and form. Action which is brought by tyrannical violence, by pushing, is associated with matter, since there must be a material connection for the action to take place. This is the kind of action shown by electrics, and we shall follow and repeat

Gilbert's beautiful experiments which are meant to show this point. But there is also action associated with form, and that is the kind of action that magnetism is. "Quitting the opinions of others," he says in Chapter Four of Book Two, "about the attraction of the lodestone, we will now show the reason of its action and the nature of its motion. There are two kinds of bodies that are seen to attract bodies by motions perceptible to our senses—electric bodies and magnetic. Electrical bodies do this by means of natural effluvia from humor; magnetic bodies by formal efficiencies or rather by primary native strength [or power]. This form is unique and peculiar." Gilbert explains that it is not the form that the Greek philosophers called the formal cause (again look back at Chapter One), nor any of the medieval varieties of this. "It is the form of the prime and principal globes; and it is of the homogeneous and not altered parts thereof, the proper entity and existence of which we may call the primary, radical and astral form." This is not Aristotle's form, "but that unique form which keeps and orders its own globe. Such form is in each globe—the Sun, the Moon, the Stars—one; in Earth also tis one, and it is that true magnetic potency which we call the primary energy." The Earth's true form is a sphere, so every bit of prime terrene matter, true Earthly matter, will tend to arrange itself into that shape by taking up the appropriate orientation of the place it finds itself. A lodestone is made up of rather weathered prime terrene matter but has enough to ensure that it takes up the orientation appropriate to being a part of a globe, the Earth's globe. And this power of controlling matter extends outside the Earth's actual globe, which is just that amount of matter that has so far taken up that form. "Hence the magnetic nature is proper to the Earth and is implanted in all its real parts according to a primal and admirable proportion. It is not derived from the heavens as a whole, neither is it generated thereby through sympathy, or influence, or other occult qualities." And here he departs radically from the magical tradition which saw the Earth as a microcosm, under the influence of the stars. The Earth has its own sphere of influence according to Gilbert, and so has every other primary globe. The forms of the Sun and Moon are operative in just the same way. "A little fragment of the Moon arranges itself, in accordance with the lunar laws so as to conform to the Moon's contour and form, or a fragment of the Sun to the contour and form of the Sun, just as a lodestone does to the Earth or to another lodestone, naturally tending towards it and soliciting it."

This is in the nature of a very general hypothesis which is not susceptible of direct proof. This idea of the form of the primal globes is a new concept, a new idea within the reach of which the actual things that happen are supposed to make sense. In a way there is no discovery that could prove this idea to be right. It is almost exactly the modern idea of the field, a spatially extended potentiality for acceleration and orientation. This is Gilbert's idea of form. But experiments can be devised to eliminate the more obvious rival ideas, since one of these at least has consequences over and above the magnetic phenomena which they are meant to explain. Suppose there were a magnetic effluvium, some stuff which is passed out from the lodestone and taken up by the iron, perhaps a stream of atoms such as in the explanation conceived by Lucretius. Now if this theory were correct it ought to be possible to detect by some sensitive device a stream of atoms, or of air, strong enough to move the iron which is attracted to the lodestone. He sets up the experiment by contrast with his experiments on electrics, which we shall describe later. " For the electric effluvia," he says, " as they are hindered by the interpolation of any dense body, so too are unable to attract through a flame or if a flame be nearby. But iron, which is hindered by no obstacle, from deriving from the lodestone force and motion, passes through the midst of the flame to join the lodestone. Take a short piece of iron wire, and when you have brought it near to a lodestone it will make its way through the flames to the stone ; and a needle turns no less rapidly, no less eagerly to the lodestone though a flame intervenes than if only air stands between ". In this way all theories which suppose a physical connection are thought by Gilbert to have been shown to be false. Magnetic coition does not occur by atomic rebound, nor by the infilling of a vacuum, nor by the extension of subtle rods, as was supposed by Cornelius Gemma. Neither is it due to the supposed fact that lodestones are alive, nor is the even more fanciful theory of Porta any better. We have already heard Gilbert's opinion of that. Finally he dismisses the traditional Four Elements theory. " As for the causes of magnetic movement, referred to in the schools of philosophers to the four elements and the prime qualities, these we leave for roaches and moths to prey upon." In short, magnetic coition " takes place in all prime bodies, and in bodies that are allied and especially those that are near akin to these, and this on account of identity . . . all true Earth substance draws its kind ". Magnetism then is not an effect depending on a material intermediary.

Gilbert develops the field idea still further, and draws an analogy with light. " The magnetic force is given out in all directions around the body . . . but the sphere of influence does not persist, nor is the force that is diffused through the air permanent or essential ; the lodestone simply excites magnetic bodies situate at a convenient distance." Here we have very nearly expressed the modern idea of potentiality. There is nothing permanently existing around a lodestone which acts upon a body that finds itself there. Instead, there is a power or potentiality to excite a magnetic body. Gilbert believes that light is a potential rather than an actual stuff. " And as light—so opticians tell us—arrives instantly, in the same way, with far greater instantaneousness, the magnetic energy is present within the limits of its forces ; and because its act is far more subtle than light, and it does not accord with non-magnetic bodies, it has no relations with air, water or other non-magnetic body ; neither does it act on magnetic bodies by means of forces that rush upon them with any motion whatever, but being present solicits bodies that are in amicable relations to itself. And as a light impinges on whatever excites it, so does the lodestone impinge upon a magnetic body and excites it." The shape of the *orbis virtutis*, or magnetic field as we should call it, is determined by the shape of the lodestone, being spherical around spherical lodestones, and having other shapes around others. Gilbert does not plot these fields very carefully. He uses a compass needle to mark out some lines of the field, and notices that the lines of force do not pass to the centre of the magnetic body, but instead go towards the magnetic poles. On p. 247 of the Mottelay *De Magnete* there is reproduced a splendid drawing of the field of a longitudinal lodestone, not unlike a bar magnet (fig. 3). It is remarkable that with all these clues in his hand, so to speak, and with the use of small pieces of iron so much a feature of experiments on the magnet, no-one seems to have seen the field as a structure of lines as Michael Faraday came to see it. If we think of Porta's magnetic army game, it is surely strange that he did not see the filings and chips of lodestone arrange themselves in those characteristic patterns which are so familiar to us.

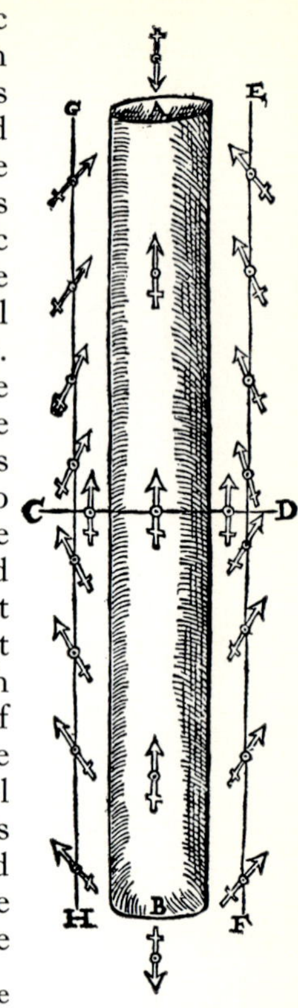

Fig. 3. The directional influence of the lodestone, shown by small compass needles.

To sum up : the magnetic effects derive from the forms of primal bodies such as the Earth or the Sun. They are directional and not attractional phenomena. The lodestone sets itself in the direction it would have as a part of a primal globe, and this is because it has some amount of primal matter in it. Each important body has a form, and hence a magnetic field or *orbis virtutis* around it. A lodestone being made in part of primal matter must therefore have a primal form of its own and is in fact a little earth.

II. *Electrics*

The clear distinction of electrical from magnetic phenomena is one of Gilbert's permanent contributions to the study of the subject. After pointing out that amber is usually found washed up on the shores of Mediterranean countries and sometimes even Britain, Gilbert goes on to argue that it is just untrue that only amber has the power of attraction and that it only attracts chaff. His first move in studying the phenomenon is to try to list as many substances as he can that show this effect, that is, have this *power*. " For not only do amber and jet, as they suppose, attract light corpuscles (substances): the same is done by diamond, sapphire, carbuncle, iris stone, opal, amethyst, vinventina, English gem (Bristol stone, *bristola*), beryl, rock crystal. Like powers of attracting are possessed by glass, especially clear and brilliant glass; by artificial gems made of (paste) glass or rock crystal, antimony glass, many fluor-spars and belemnites. Sulphur also attracts, and likewise mastick, and sealing wax (of lac), hard resin, orpiment (weakly). Feeble power of attraction is also possessed in favouring dry atmosphere by sal gemma, mica, rock alum." Gilbert has an asterisk beside these substances in his book, and this means that he knew the facts about them from experiment. What he has discovered about them is that they all behave in a certain way, namely that when rubbed they will attract small bits of various stuffs. He now knows that they all possess a certain power, that is a certain capacity to behave in a certain way in certain circumstances. But that is not the end of a scientific investigation. If it were he could have stopped his study of electrics almost at that point. He also needs to know *why* these, and these substances alone, have the power. He needs to form some idea of their *nature*, and to see if there is something about their nature which they have in common and which substances which lack this power also lack.

Before that question is pursued some preliminaries must be done. By another series of experiments Gilbert showed that it was also untrue that only chaff was attracted to electrics. " These several bodies," he notes, " not only draw to themselves straws and chaff, but all metals, wood, leaves, stones, earths, even water and oil; in short whatever things appeal to our senses or are solid."

Having settled the question of the universality of the effect, Gilbert provides himself with a new instrument, invented by himself. This is the *versorium*, what we call a field indicator. " Now in order clearly to understand by experience how such attraction takes place, and what those substances may be that so attract other bodies, and in the case of many of these electrical substances, though the bodies influenced by them lean toward them, yet because of the feebleness of the attraction they are not drawn clean up to them, but are easily made to rise, make yourself a rotating needle (electroscope—*versorium*) of any sort of metal, three or four fingers long, pretty light, and poised on a sharp point after the manner of a magnetic pointer. Bring near to one end of it a piece of amber or a gem, lightly rubbed, polished and shining: at once the instrument revolves."

Just as in the case of magnetism, so for electrics, there is a variety of traditional theories of which Gilbert has a poor opinion. He easily disposes of the theory that it is heat which is responsible for the attraction, by heating some electrics and finding that they do not attract the versorium, unless that heating is produced by friction or rubbing. So it is the rubbing and not the heat that is responsible. Nor is the effect due to the inrush of air, since a candle flame which consumes air and draws it towards it does not attract. And finally it follows from the variety of electrics that it cannot be some special property of amber that gives it the power. And there is another class of substances that even when rubbed do not attract, and so must be grouped as non-electrics. These are " emerald, agate, carnelian, pearls, jasper, chalcedony, alabaster, porphyry, coral, the marbles, lapis lydius (touchstone, basanite), flint, bloodstone, emery or corundum, bone, ivory ; the hardest woods, as ebony ; some other woods, as cedar, juniper, cypress ; metals, as silver, gold, copper, iron. The lodestone, though it is susceptible of a very high polish, has not the electric attraction ". Gilbert notices in this chapter how dependent on the weather conditions such experiments are, and as he says : " This we may observe," speaking of the attraction of electrics, " when in mid-winter the atmosphere is very cold, clear and thin ; when the electric effluvia of the Earth offer less impediment, and electric bodies are harder ".

But what is the nature of electric bodies in virtue of which they exhibit electric behaviour when rubbed ? Gilbert points out that the materials of the Earth are of two kinds, watery and humid, and firm and dry. " From this two fold matter, or from the simple concretion of one of these matters, come all the bodies around us, which consist in major proportion now of terrene matter, anon of watery." He has already pointed out that amber must at one time have been more fluid than it now is, since there are often insects trapped in it. Perhaps using this clue he draws the conclusion : " those bodies that derive their growth mainly from humours, whether watery humour or one more dense ; or that are fashioned from these humours by simple concretion, or that were concreted out of them long ages ago ; if they possess sufficient firmness, and after being polished are rubbed, and shine after friction,—such substances attract all bodies presented to them in the air, unless the said bodies be too heavy ". But why are those materials formed from concreted watery substance capable of being electrified ? As Gilbert puts it : " And now, at last, we have to see why corpuscles are drawn toward substances that derive their origin from water, and by what manner of force, by what hands, so to speak, such substances lay hold of matters nigh them ".

We must remind ourselves that Gilbert considers that the way electrics act is by a material connection, and actual stuff comes out of the electric and forms a material connection with the particle that is drawn. It is a much feebler effect than magnetic effects, and very easily shielded or destroyed, even by just blowing on the amber. " And now what is it that produced the movement ? The body itself circumscribed by its contour ? Or is it something imperceptible for us flowing out of the substance into the ambient air ? " Gilbert asks whether, if it is an effluvium that is responsible, something flowing out of the electric, it acts upon the air and the air causes the corpuscles attracted to move, or whether it is a direct action of the effluvium. It cannot be the lustre or shininess of the gems or amber because some shiny things do not attract at all. He notes that the electric effluvia must be much more tenuous than the fire-producing effluvia struck off from flint, since the electrical ones will not take fire. "They are not a breath, for, when given forth, they do not exert propelling force; they flow forth without any perceptible resistance, and reach bodies. They are exceedingly attentuated humours, much more rarefied than the ambient air; to produce them requires bodies generated of humor and consolidated to considerable hardness. Non-electric bodies are not resolvable into humid effluvia."

To decide whether the amber attracts the air or attracts the body itself Gilbert offers a neat little experiment. "It plainly attracts the body itself in the case of a spherical drop of water standing on a dry surface; for a piece of amber held at a suitable distance pulls toward itself the nearest particles and draws them up into a cone; were they drawn by the air the whole drop would come towards the amber." He also tries the same proof by bringing a piece of amber near a small flame, and shows that the flame is not drawn by the amber.

The upshot of this line of reasoning and experiment is this: "A breath, then, proceeding from a body that is a concretion of moisture or aqueous fluid, reaches the body that is to be attracted, and as soon as it is reached it is united to the attracting electric; and a body in touch with another body by the peculiar radiation of the effluvia makes the two one: united, the two come into most intimate harmony, and that is what is meant by attraction." An electric attracts because it exudes a fluid which encompasses the body to be attracted. So far the only reasons offered for this theory are the refutations of some of its more obvious rivals, and some rather dubious stuff about the origins of electrics. Nowhere is Gilbert's complete grasp of scientific method so beautifully shown as in his next move in this study. The effluvium is *like* a fluid, so let us experiment with actual fluids to see if they do behave in such a way as to make the analogy upon which the concept of effluvium depends a viable one. The effluvium theory offers a hypothetical mechanism of the electric attraction. It is based on a fluid analogy. To see if it is reasonable we must see if the behaviour of fluids *upon* which the effluvium theory is modelled is such as to include attraction and repulsion phenomena.

"Bodies tend to come together and move about on the surface of water like the rod C (see fig. 4), which dips a little into the water. Evidently the rod EF, floated by the cork H and having only the wetted end F above the water's surface, will be attracted by the rod C, if C be wetted a little above the water's surface. . . . A wet object on the surface of water seeks union with another wet object when the surface of the water rises in both. . . . But if the whole rod C be dry above the water, it no longer attracts but repels the rod EF." Evidently attractive phenomena are shown by wet and dry things. Now Gilbert draws out the analogy. " So do bodies concreted from liquids when melted a little in the air exercise attraction, their effluvia being the means of unition ; for the water in humid bodies or in bodies drenched with superficial moisture on the top of water has the force of an effluvium."

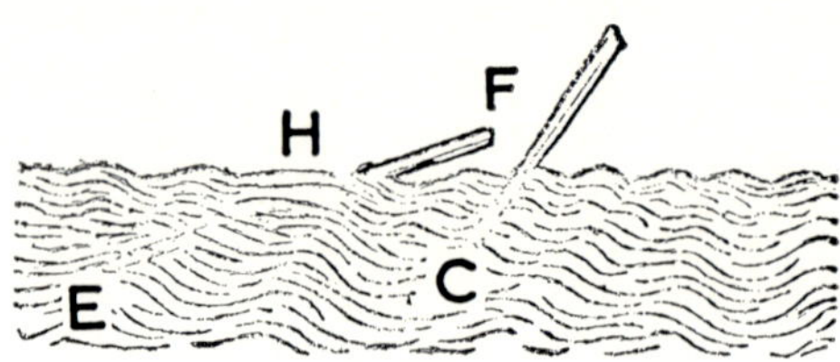

Fig. 4. The floating rod EF is "attracted" by the rod C.

Gilbert sums up the result of these researches into the causes of the behaviour of electrics and magnets in the following words : " The difference (distinction) between electric and magnetic bodies is this : all magnetic bodies come together by their joint force (mutual strength) ; electric bodies attract the electric only, and the body attracted undergoes no modification through its own native force, but is drawn freely under impulsion. . . ." That is a body attracted by an electric does not, Gilbert believes, itself become an electric, in great contrast to the magnetic bodies. " The matter of the earth's globe is brought together and held together by itself electrically. The Earth's globe is directed and revolves magnetically ; it both coheres and, to the end it may be solid, is in its interior fast joined." In short " The electric motion is the motion of coacervation of matter ; the magnetic is that of arrangement and order."

IV. *The properties of magnets*

Gilbert repeats with more care the experiments of his predecessors, but he extends his investigations a good deal further. One of his most striking innovations was his reform of the nomenclature of the poles. Up until his time it was the custom to speak of the pole which turned to the north as the north pole of the magnetic body, and the pole which regarded the south part of the Earth as the south pole of the magnetic body. If, to anticipate the argument, we suppose that the earth is a magnet itself, then it will have poles and these are north at its north, and south at its south. So when the magnetic body orients itself if suspended freely, the pole which has the same nature as the Earth's south pole will regard the north, and the pole which has the same nature as the Earth's north pole will regard the south. The pole of the magnetic body which points to the north is a south pole, while the pole which points to the south is a north pole. Thus for Gilbert a north magnetic pole is that kind of pole that is found at the geographical north of the Earth. He notices that if a piece of a magnet is broken off, because it acquires opposite verticity i.e. power of " regarding " a particular pole of the Earth to the body from which it is broken, it will not turn around when freely suspended but remain in the orientation in which it was when a part of the original body.

The bulk of Gilbert's experiments that are concerned with the lodestone as such, and not with its relations with the Earth, is concerned with the lodestone and iron. He was particularly interested in those effects which are brought about by a combination of iron and lodestone, and performed a great many experiments on lodestones with iron caps on them, called " armed " stones One of the main results to emerge from these studies is the difference between the unition of magnetic bodies and their coition. The power of unition is the power to lift certain weights, and the greater the unition the greater the weight that can be lifted. The power of coition is the power of mutual influence between bodies, and is treated by Gilbert as indicated by the distance away that a piece of iron can be and still be drawn by the stone.

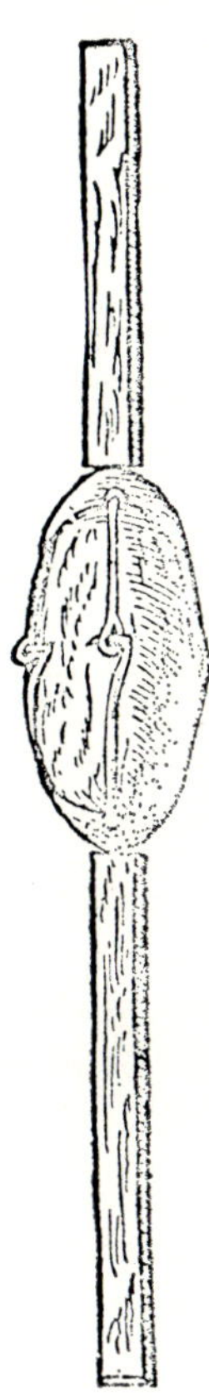

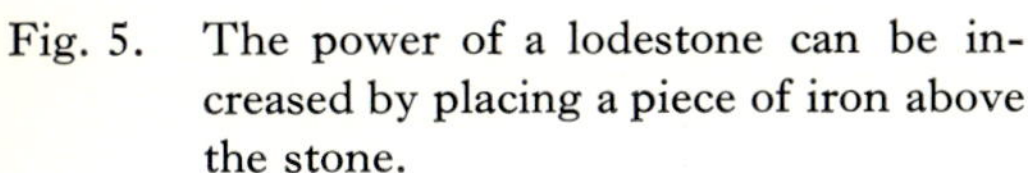

Fig. 5. The power of a lodestone can be increased by placing a piece of iron above the stone.

He shows that though an armed and an unarmed lodestone can draw a particular piece of iron from the same distance, the armed stone has the greater lifting power. He combines this with another important insight, namely that when two magnetic bodies cling together they form one magnetic body, and since he believes that lifting power is roughly proportional to the weight of a lodestone, the single body formed by two bodies clinging together should be able to lift greater weights than either can lift separately. He describes an ingenious experiment designed to show this. If one takes a lodestone, and holds it with the long axis vertical, as in fig. 5, then it will lift the piece of iron C, only if it is surmounted by another piece of iron A, and not if that piece of iron is removed. These investigations are really very much subordinate in the general plan of Gilbert's researches. What we have in fact is a step by step attempt to establish nothing less than a magnetic cosmology The first step is made in the great Gilbertian hypothesis that the Earth is itself a lodestone. It is upon this hypothesis that the *De Magnete* is built.

V. *The Earth is a lodestone*

Gilbert begins his exposition of this idea by pointing out how little of the Earth we really know. There are shallow seas, not much more than a mile (1500 metres) deep, and there are mines, no more than 500 fathoms (1000 metres) deep. The surface of the Earth is composed of very various substances, soils of various kinds, and minerals and metals. All this variety has been produced by the effects of weather and of the influences of the heavenly bodies upon the prime terrene matter, the real stuff of the Earth. But we never have a chance to come across that. We cannot mine down to the centre of the Earth, deterred as we are " by the enormous cost of executing such vast undertakings ". However, the materials on the surface, though weathered, do resemble in some degree " their source, because their matter is of the earth, albeit they have lost the prime qualities and the true nature of terrene matter ". But of all these the lodestone is most like the Earth's true matter. " All magnetic bodies ", Gilbert says, " seem to contain within themselves the potency of the earth's core and of its innermost viscera, and to have and comprise whatever in the Earth's substance is privy and inward : the lodestone possesses the actions peculiar to the globe, of attraction, polarity, revolution, of taking position in the Universe according to the law of the whole ; it contains the supreme excellencies of the globe and orders them : all this is token and proof of a certain eminent combination and of a most accordant nature." The lodestone, when spherical has all the properties of the Earth, a natural equator, natural poles, it draws bodies to it that conform to its geometry, and it is itself such as to confoım to the geometry of the Earth. A strong lodestone, Gilbert believes, possesses the prime telluric form ; it is as near as we can find a piece of the real stuff of the Earth. No wonder then that it orients itself to conform to the form of the Earth, when it is within the *orbis virtutis*. Not only that, but Gilbert also thinks that all hard rocks when freed from the detritus within which they lie and with which they are contaminated, will be capable of being magnetized and then show their true terrestrial nature by orienting themselves according the form of the Earth. Thus the Earth and the lodestone are identical. From this follows a consequence of very great practical importance. *For experimental purposes a spherical lodestone will do just as well as the whole Earth.* Many of Gilbert's most important experiments depend upon and develop the analogy between a spherical lodestone and the Earth. This is why he speaks of the spherical lodestone as a *terrella*, or little Earth. But he thinks that the Earth is not just like a lodestone, it is one. So whatever can be demonstrated on the lodestone must be true of Earth also. Thus the subject of experimental magnetic geography is broached.

By experimenting with his terrella Gilbert establishes really two main propositions about the magnetic geography of the Earth.

Proposition one. Some of the geographers' concepts are real, natural features of the Earth, and some are not. The real ones are the poles and the equator, since they are distinguishable by magnetic experiments. At the poles the magnetic force is at its strongest, for there all the various forces of the parts of the lodestone come together and are concentrated. There too a small piece of iron wire stands vertically upright upon the surface of the terrella. But at the equator the strength is weakest, and a small piece of iron wire lies horizontally upon the surface. The magnetic meridians are real too, since they are the representatives of the directional power of the form of the lodestone, passing through the poles and crossing the equator. The tropics of Cancer and Capricorn, and the Arctic and Antarctic circles are not real, they are introduced solely for the convenience of mathematicians, in identifying places on the Earth, or charting the stars on the celestial sphere.

Proposition two. The " formal energy " of the lodestone does not reach out from every point on the surface at the same angle. This is the phenomenon that we recognize as the directional changes of the lines of the magnetic field. Gilbert gives a geometrical construction for finding the direction of magnetic orientation. This is reproduced in fig. 6. E is the north magnetic pole, and HQ is the equator. Suppose we are trying to find the contribution of the magnetic power at A. Then we find, according to Gilbert, that it contributes towards the power at C, F, N and E, but not to any point between C and H. If we want to find the direction of the magnetic power at C, say, then we draw a chord from H to C, and that is the direction. These geometrical constructions are not, of course, correct. It is scarcely surprising that Galileo was not enthusiastic about Gilbert's geometrical talents. Nevertheless Gilbert was right in supposing that the angle of the magnetic direction changes in some complex way between the horizontal at the equator, and the vertical at the poles. This proposition is, in effect, an effort to express, in mathematical terms, Robert Norman's discovery of the fact of dip (or declination as he called it).

To each of these propositions and their development, Gilbert devotes a complete book of the *De Magnete.* To the phenomenon of direction, the set of the magnetic needle along the meridians, he devotes Book III, and to the phenomenon of dip he devotes Book V. In order to apply the principles of magnetic geography to the actual world one must take account of various small and irregular variations of the actual direction in which the needle sets, from the magnetic meridian, and so from true north. To these variations Gilbert devotes Book IV of the *De Magnete*. In order to understand these books thoroughly we must

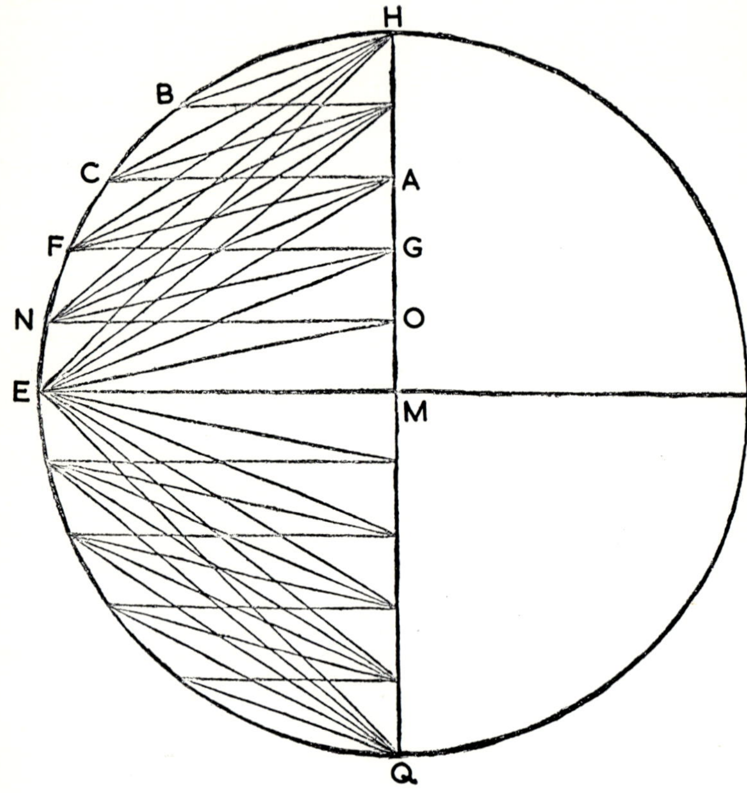

Fig 6.

be aware of a spectacular error into which Gilbert's very concepts drew him. *He did not, because he could not, distinguish the geographical from the magnetic poles*. To our way of thinking the geographical poles are the points where the Earth's axis of rotation cuts the surface of the Earth's sphere. And for us the magnetic poles are the points where the dip needle sets vertically. To us it is a matter of fact whether the geographical pole and the magnetic pole coincide, and for us it is a matter of fact that they do not. Remember that for Gilbert the Earth *is* a lodestone, in its primary nature and essence, so that all the terrestrial phenomena must be referred to its magnetic features, including its rotation. The poles are the magnetic poles, and the equator is the magnetic equator. The true magnetic meridians cannot fail to pass through the magnetic pole and since that can be the only pole, through the geographical pole too. Any tendency for the needle to point anywhere but true geographical north, must, for Gilbert, indicate a variation, and not a deviation between the magnetic meridian and the geographical one. So he could not have discovered that the geographical and magnetic poles differed in location, his conceptual system precluded it. In speaking of the subject matter of Book III, *Direction*, he says : " This discrepance is known as the variation of the needle and of the lodestone ; and as it is produced by other causes and is, as it were, a sort of perturbation and deprivation of the true direction, we propose to treat here only of the true direction of the compass and the magnetic needle, which would all over the earth be the same, toward the true poles and in the true meridian, were not hindrances and disturbing causes present to prevent ".

The main points made in Gilbert's study of direction have to do with the nature of poles. The verticity of a lodestone or a piece of iron is its polarity and directedness. The orientation of the lodestone or magnetic needle is not due to attraction between the poles but to verticity, to the directional power of the terrella or the Earth. The poles are the places of greatest force because there the magnetic energy is concentrated, but the directional power is a property of the whole object, be it Earth or terrella, and is in fact, as we have had explained earlier, the form of the object which has the magnetic power. Direction then, is nothing but the manifestation of form in orienting magnetic bodies. The main run of experiments described in Book III has to do with the changes in verticity brought about by various operations upon lodestones and iron.

These fall into two main groups. He studies the effects upon iron objects of rubbing with lodestones, and of heating, and so on. And confirms the well-known ideas that heating destroys verticity, and that the verticity can be reversed by rubbing an iron object with the opposite end of the lodestone from which it originally acquired verticity. He notices too that, in conformity with his theory that the magnetic power is a function of the whole of a magnetic body, heating a small part of magnetized iron wire does not destory its magnetism, but putting the whole wire in the fire does.

The other group of experiments studies the changes in verticity brought about by dividing lodestones. Some of the facts he establishes had been known already, but in these experiments he carefully investigates most of the possibilities, noticing the formation of new poles when lodestones are broken in two, and determining which poles are which after breaking. Again he tries to make rough quantitative judgments, pointing out that the power of the pieces of a lodestone are roughly proportional to their weights.

VI. *Variation*

In his experimental study of variation Gilbert does really break quite new ground. Without the idea of the spherical lodestone as a model of the Earth, a terrella, this set of experiments could not have been conceived. By using the terrella and matching it more and more to the actual Earth, Gilbert tried to determine the causes of the variation from true north of the magnetic needle. In these studies two sets of empirical facts bear upon each other. From discussions with Drake and others Gilbert had a fair knowledge of the behaviour of the compass in a wide variety of places, both in the Southern Hemisphere and on both sides of the Atlantic. His experiments were designed to bring into conformity with these facts the results of experiments with carefully modified terrellas. In these studies we have one of the classical uses of a *model.* Gilbert was undoubtedly good at models. In his studies of electricity we have seen his beautiful employment of a fluid model to explain the unknown mechanism of electrical attraction, and his efforts to use the model itself as an experimental set-up. Here the model is the product of analogy between quite different sorts of phenomena, between electrical effects and fluids. Such a model is called a *paramorph.* But Gilbert believed that his terrella and the Earth had a stronger relationship since both, he thought, were *really* lodestones. The terrella was literally a " little Earth ", a scale model of the Earth, whereas the wet and dry wires are not literally electrically charged, they just behave in some respects like electrics. Scale models are one of the kind of model called *homoeomorphs.* His studies of variation led him to carve terrellas into more and more accurate scale models, or homoeomorphs, of the real Earth. Upon their surfaces he placed compasses and tried to find in this way the causes of variation.

Gilbert believed that the cause of variation was the actual deformity of the Earth's true shape due to unevenness in its surface. " But as the globe of the earth is at its surface broken and uneven, marred by matters of diverse nature, and hath elevated and convex parts that rise to the height of some miles and that are uniform neither in matter nor in constitution but opposite and different, it comes about that this entire earth-energy turns magnetic bodies at its periphery toward stronger massive magnetic parts that are more powerful and that stand above the general level. Wherefore at the outermost superficies of the Earth magnetic bodies are turned a little away from the true meridian. And since the Earth's surface is diversified by elevations of land and depths of seas, great continental lands, oceans and seas differing in every way,—while the force that produces all magnetic movements comes from the constant magnetic Earth-substance, which is strongest in the most massive continent and not where the surface is water or fluid or unsettled,—it follows that toward a massive body of land or continent rising to some height in any meridian (passing whether through islands or seas) there is a measurable magnetic leaning from the true pole toward east or west, i.e. toward the more powerful or higher or more elevated magnetic part of the Earth's globe."

So far we have a hypothesis only. But Gilbert is a scientist and he tells us that " we have therefore to inquire how the demonstration of this new natural philosophy may be drawn from unquestionable experiments ". What we have to prove is that " variation is due to inequality among the earth's elevations ". But how to do it ? Fortunately our scale model is to hand, and " this very thing is clearly demonstrated on the terrella thus : " a lodestone with a large depression in it can serve as a model to establish the general point, before we proceed to make accurate scale models of the Earth, with continents and all. Variation is clearly demonstrated by putting compass needles in the depression when they clearly vary from meridian towards the elevated part which they are nearest. Right in the centre of the depression, equidistant from the elevated sides, the needle still lies along the meridian. Again using a terrella with projecting lumps it is easily demonstrated that the compass needles vary from the meridian towards the lumps. " The demonstration is made with small bars or short needles placed on the terrella : they turn from the terrella toward the projecting mass and the great eminences."

The interpretation of these results is important and interesting. Having noted that variation is constant at a given place and that islands do not produce variation, Gilbert explains the phenomena in terms of his general theory. Since it is because of a tendency to set in the direction appropriate to the true form of the Earth that the compass has a directional character, then it will show this whatever form the true stuff of the Earth has. In the great continents the prime terrene matter is raised up and there the Earth's form itself departs from a perfect sphere. It is *not* that the continents attract the needle this way and that, but that the needle conforms to the true form of the Earth, and it varies because that form varies from a perfect sphere. For this reason a little island, which can be presumed to contain none of the Earth's real terrene matter does not seem to cause any variation.

Porta and others had supposed that the longitude could be found by accurate determinations of variation, since they supposed that the variation was proportional to the longitude. This hypothesis prompts Gilbert to make a general survey of the known facts about variation in different parts of the Earth, from which it clearly emerges that it is " in no wise true ". The variation does not increase to the east as the compass is taken eastwards. Gilbert realized that the variation was greatest in the polar regions but failed to draw the conclusions that showed that the magnetic and geographical poles might diverge. Since this conceptual system could not allow that these concepts were not identical the polar variations must show distortions in the form of the Earth at that place. Furthermore, he supposed that the variation in the centre of oceans and in the heart of continents was nil. This accorded well with his general theory, and it would also follow that as the compass was carried towards a continent from the centre of an ocean the variation would increase, since it would become sensitive to the changed and varied configuration of the prime terrene matter raised in the continent.

VII. *Dip*

The phenomenon which Robert Norman called declination, Gilbert's translator renders by our modern term " dip ". " We come at last," says Gilbert, " to that fine experiment, that wonderful movement of magnetic bodies as they dip beneath the horizon in virtue of their natural verticity."

Robert Norman, who had discovered dip, had described the dip-circle and Gilbert adds little to the instrument. However, Norman had merely conjectured that the dip would vary with latitude, and had had no clear idea of how this would happen. Here Gilbert had a tremendous advantage, because he had his model Earth, his terrella to work with. Upon the surface of the terrella all the magnetic phenomena of the Earth would be displayed. By this means he readily establishes the general principles of the dip phenomenon on the Earth. At the poles dip is 90° and at the equator, dip is zero. But between the poles and the equator there is a continuous variation in dip, and it is to the law of this variation that he devotes great attention, since if it could be discovered there would immediately be possible a determination of latitude. With a spherical lodestone the variation in dip is easily shown, " by means of a number of bits of iron wire of equal size, one barley-corn in length, and placed in a meridian. At the equator the bits of iron are directed towards the poles, and lie upon the body of the terrella in the plane of its horizon. The nearer they are placed to the poles the more do they rise from the horizontal by reason of their turning poleward ; at the poles they tend straight to the centre. But bits of iron will not stand upright, save on a good lodestone, if they be too long ".

Like Norman, Gilbert rejects the idea that dip is caused by the attraction of the lodestone, and he repeats Norman's beautiful experiment with the goblet of glass and the floating compass needle which we have noticed already. But with his own theory of magnetism to hand he does not need to remain satisfied merely with the idea of a " point respective ", as Norman had done. " Throughout nature," he says, " we have to recognize that wondrous work of the Maker whereby the principal bodies are restricted within particular localities and, as it were, hedged round with fences, nature so ordering." It is of course the form of the lodestone which determines that the needle should set in the angle of dip. And this he shows by a variety of experiments, the most convincing perhaps being the refutation of the attraction theory by showing that dip is the same whatever may be the strength of the lodestone. If it were a phenomenon due to the attraction of the lodestone, then it would

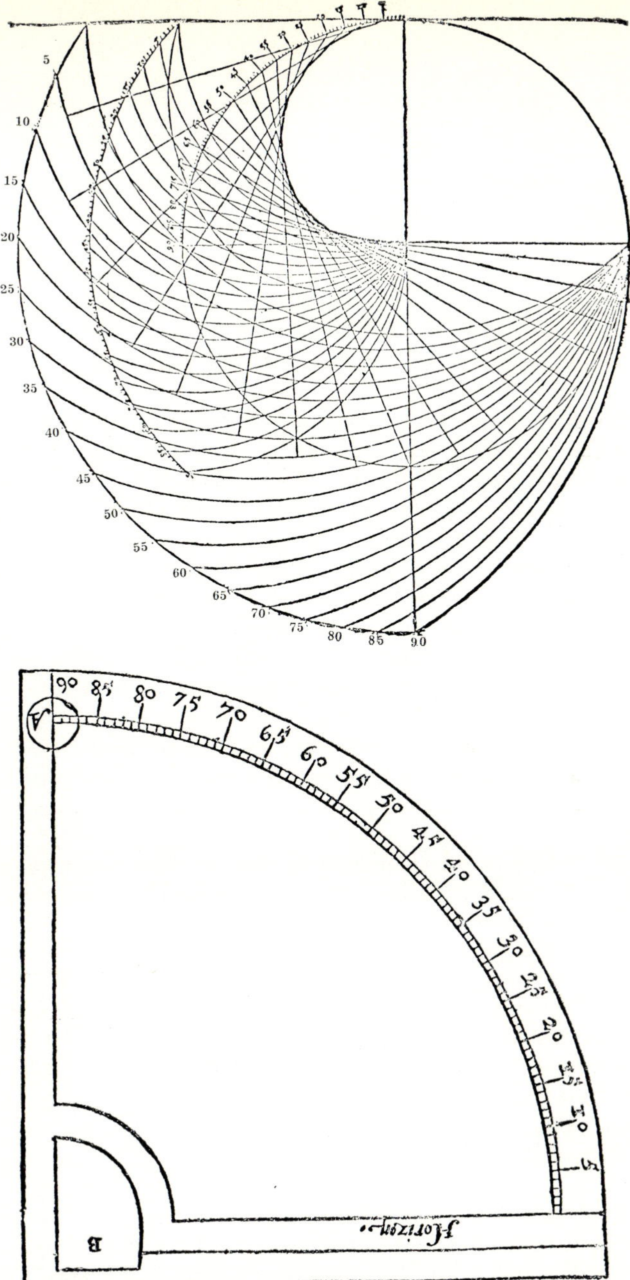

Fig. 7. The graduated quadrant is placed over the engraved plate, and set as described in the text.

be reasonable to suppose that the stronger the lodestone, the larger would be the angle, but this is not so. " The versorium simply revolves so far as nature demands."

The critical practical question is the relation of dip to latitude. This is an acute problem since it dips " not in ratio to the number of degrees or over a like arc ; but over a very different one, for this movement is in truth not a dipping movement, but really a revolution movement, and it describes an arc of revolution proportioned to the arc of latitude ". In short the versorium does not regard the centre of the lodestone, since, if it did, then dip would be proportional directly to latitude. Instead, " it obeys the whole and its mass and outer limits, the powers of both co-operating, to wit, those of the magnetized versorium and of the earth ". In Chapter Seven of Book V an elaborate construction is given for finding the angle of dip, and in Chapter Eight it is elaborated into a full-scale computing device for connecting latitude with dip. This provides a method " how to acertain the elevation of the pole, or the latitude of any place, by means of the following diagram, turned into a magnetic instrument, in any part of the world, without the help of the heavenly bodies, Sun, planets, or fixed stars, and in foggy weather as well as in darkness." With what a paean of joy Gilbert greets this device. " We can see how far from idle is the magnetic philosophy ; on the contarary, how delightful, how beneficial, how divine ! Seamen tossed by the waves and vexed with incessant storms, while they cannot learn even from the heavenly luminaries aught as to where on Earth they are, may with the greatest ease gain comfort from an insignificant instrument, and ascertain the latitude of the place where they happen to be." Using the instrument drawn out in fig. 7 they can use their observation of the angle of dip to find the latitude, by noting the degree of dip " on the inside arc of the quadrant, and the quadrant is turned round at the centre of the instrument until that degree on the quadrant touches the spiral line : then in the open space B, at the centre of the quadrant, the latitude of the region on the periphery of the globe is found by the *linea fiduciae AB* ". The quadrant should be pivoted at the point A, and fixed to the centre of the circle in fig. 7. Gilbert gives some practical instructions about using the dip instrument at sea, keeping it vertical and its base horizontal, and in the line of a meridian. He was also aware of variation of dip, and offers the same explanation of that phenomenon as of the variation of direction, namely that it is due to the needle following deformations in the true underlying terrene matter of the earth.

VIII. *Magnetic cosmology*

For Gilbert magnetism is the most fundamental phenomenon of nature since it is the exhibition of a power that derives from the very form of the principal bodies of the Universe. It must, therefore, be capable of cosmological application, and the general structure of the

Universe must derive from it. It is in his magnetic cosmology that we find Gilbert's closest connection with the tradition of the great magicians and of Hermes Trismegistus. The Universe was supposed to be animate, and magnetism is its soul. " Wonderful is the lodestone shown in many experiments to be, and, as it were, animate ", Gilbert begins cautiously. But he goes on : " And this one eminent property of the same which the ancients held to be a soul in the heavens, in the globes, and in the stars, in Sun and Moon. For they deemed that not without a divine and animate nature could movements so diverse be produced, and such vast bodies revolve in fixed times, or potencies so wonderful be infused into other bodies ; whereby the whole world blooms with most beautiful diversity through this primary form of the globes themselves." However some did not feel this, for example " Aristotle held that not the Universe is animate, but the Heavens only". This view excites Gilbert's scorn, since it implies that " in comparison with the whole creation " the Earth " is a mere mite, and amid the mighty host of many thousands is lowly, of small account, and deformate ". And the upshot of adopting this point of view is that " Aristotle's world would seem to be a monstrous creation, in which all things are perfect, vigorous, animate, while the earth alone, luckless small fraction, is imperfect, dead, inanimate and subject to decay ". The ancient magicians were better informed. Hermes, Zoroaster and Orpheus thought that there was a universal soul. Gilbert gives it as his own opinion that " the whole world [is] animate, and all globes, all stars, and this glorious earth, too, we hold to be from the beginning by their own destinate souls governed and from them also to have the impulse of self-preservation ".

The form which manifests itself as magnetism is the soul of the animate globes, of which the Earth is one. The Earth can act, through this form, and it can act for its own self-preservation. To those who might object that the Earth cannot be animate since it has no organs, Gilbert replies that in some plants it is hard to make out any organs, and visible organs do not seem to be needed for things to be alive. We cannot distinguish organs in the stars and planets. Who can deny that God is soul, but who would ascribe organs to Him ?

The Earth is a living lodestone. And all the astronomical facts about it derive from this. It maintains itself in position with respect to the fixed stars and the Sun *for its own benefit*, and for the benefit of the creatures which live upon it. It is thus that Gilbert explains daily rotation. The Earth naturally turns and revolves in a circular motion around its axis, because that is the way lodestones and other magnetized bodies move when displaced, for example, from their correct orientation. If the Earth were not to rotate but always to present the same face to the Sun, then that side would be overheated and burnt up and the other side would be frozen totally. " And as the Earth herself cannot endure so pitiable and so horrid a state of things on either side, with her astral magnetic mind she moves in a circle, to the end there may be, by unceasing change of light, a perpetual vicissitude, heat and cold, rise and decline, day and night, morn and even, noonday and deep night." And similarly, unless the Earth rotated daily " from the Moon also great dangers would threaten ", namely destructive tides. In a similar way the inclination of the Earth's axis out of the plane of its orbit is explained by Gilbert by the necessity to have the sequence of the seasons for the benefit of the creatures and plants on the Earth's surface. But not only that, the precession of the equinoxes can also be explained, since it is caused by a change in the inclination of the Earth's axis and so allows the influence of the stars to be differently felt, and through their astrological influence bring about those changes in the human condition which we recognize as history. Thus does a sensitive and animate Earth look after her own interests and ours.

But what about those who deny this rotation ? Basically Gilbert's argument is that " it is more accordant to reason that the one small body, the earth, should make a daily revolution than that the whole Universe should be whirled around us ". Most of the opposition, he believes, comes from a false idea about the relation of the Earth to the things on it, and to the atmosphere and the sphere of magnetic virtue. The point is that " since it revolves in a space void of bodies, the incorporeal aether, all atmosphere, all emanations of land and water, all clouds and suspended meteors, rotate with the globe : the space above the Earth's exhalations is a vacuum ; in passing through vacuum even the lightest bodies and those of least coherence are neither hindered nor broken up ".

To sum it all up Gilbert says : " By the wonderful wisdom of the Creator, therefore, forces were implanted in the Earth, forces primarily animate, to the end the globe might, with steadfastness, take direction, and that the poles might be opposite, so that on them, as at the extremities of an axis, the movement of diurnal rotation might be performed ".

Further reading

A summary of Gilbert's life and work is to be found in R. Harré, *Early Seventeenth Century Scientists*, Pergamon Press, Oxford, 1965. *The De Magnete of William Gilbert* by Duane H. D. Roller, Amsterdam, 1959 is an excellent treatment of Gilbert's book but may be hard to obtain. The *De Magnete* itself, which should be read in conjuntion with this book, is now obtainable in paperback, No. S470 by Dover Books.

Practical work

In repeating the experiments mentioned in this chapter recourse ought to be had to the *De Magnete* itself where a more detailed account of the reasons for, and point of the experiments is to be found.

Lodestones can be obtained from Gregory Botley & Co., 30 Old

Church Street, Chelsea. They are of magnetite, and can be had as small pieces of 2 or 3 cm in length suitable for directional experiments. Such a piece costs about 40 p. For terrellas, larger pieces can be obtained measuring perhaps 10 cm in diameter, but these are much more expensive, costing up to £3. However, it is not essential to have lodestones to perform the experiments. An iron ball can be magnetized and can be quite an adequate substitute for the terrella in doing magnetic geography, and bar magnets can be substituted for lodestones in many of the other experiments, if not in all.

Experiments from De Magnete

1. To show that magnetism is not due to an effluvium. Using a candle flame or bunsen burner flame, (*a*) show that a small piece of iron can be drawn through it by a magnet or lodestone ; (*b*) show that if a magnet is placed on one side of a flame then it will affect a compass needle placed on the other side.

2. The *Orbis Virtutis* of a longitudinal lodestone. By marking the direction of a small compass needle at different points along both sides of a bar magnet establish (*a*) that the directional forces " respect " the ends of the magnet, (*b*) that they do not " respect " the geometrical centre of the bar.

3. To establish the variety of electrics. By rubbing as many specimens as you can collect of the substances mentioned on p. 36, test Gilbert's hypothesis that they are all electrics.

4. Using a piece of amber or jet, or failing that a piece of the hard plastic such as pocket combs are made of, or polythene, or perspex, show that small pieces of practically any substance are attracted by the electric after it has been rubbed.

5. Following Gilbert's instructions on p. 37, make a versorium.

6. Use it in the following experiments : (*a*) that only when made warm by rubbing does amber attract, can be shown by trying rubbed amber near the versorium and then trying amber which has been *gently* warmed near a flame. (*b*) Test whether a candle flame attracts the versorium. (*c*) Try rubbing marble (or any of the other non-electrics mentioned by Gilbert on p. 37) and see if his result is correct.

7. To test whether the material or the air is drawn to the electric : bring a rubbed electric close to a drop of water. See what happens. Is it as Gilbert describes it ?

8. Push a small piece of wire through a piece of cork, and float it on the surface of some water, taking care that the wire has been wetted. Try bringing up a wet rod, just in the water surrounding and compare the effect with what happend when you bring up a dry rod, just touching the surface.

9. Hang a piece of lodestone or a length of magnetized iron from a thread, and wait until it has settled down. Which is the north pole of the *magnet* ?

10. Having found out the polarity of two magnets, see what happens when like poles are brought together, and what happens when unlike poles come together.

11. Testing the power of the armed lodestone or magnet : take two pieces of iron of roughly equal size, so that one of them is just too heavy to be lifted by a bar magnet. Place the other piece of iron on the top of the bar magnet and see if Gilbert was right in supposing that the first piece of iron can now be lifted.

12. With a terrella (either a spherical lodestone or a magnetized ball of iron) explore the magnetic geography of the Earth. First identify the poles with two small pieces of iron wire, then the equator. Mark these geographical features in with paint. Mark in a magnetic meridian. Using small pieces of iron wire show the change of dip between equator and pole. (See Frontispiece)

13. Using a dip circle, made either to Norman's or Gilbert's instructions, find the dip where you are. If you can copy the instrument depicted in fig. 7 and described in the text, use it to compare the known latitude of your town with the result got from using the table to determine the latitude from the angle of dip. If there is any discrepancy why do you think it is ?

Further reading

The subsequent history of magnetism can be followed in Charles Singer, *A Short History of Scientific Ideas*, Oxford 1959, pp. 324–5 and pp. 358–67. Modern ideas on the magnetism of the Earth can be found in *Scientific Thought, 1900–1960*, edited by R. Harré, Oxford 1969, Chapter 4.

CHAPTER 5

the origins of the knowledge of plants

THE history of our knowledge of plants shows much the same course as the history of other branches of natural knowledge. Magnificent work was done amongst the Greeks of classical antiquity. This became incorporated in a purely literary tradition in the Middle Ages, when virtually no new botanical knowledge was added, and the activities of the students of plant life were virtually confined to the copying and re-copying of the works of two or three great botanists of antiquity. The illustrations became debased and unrecognizable, the texts became corrupt, and glossaries of botanical terms lost all practical value as the languages for which they were intended disappeared from use. Paradoxically it was the sheer badness of medieval botany that led to its further development. Botanical books were meant to be used, to enable practical men such as druggists to recognize important plants. This need led to a gradual restoration of the art of botanical drawing in the fifteenth and sixteenth centuries.

Botanical knowledge was expressed in books called by scholars " herbals ". These contained descriptions and drawings of plants, with accounts of their practical uses in cooking and particularly in medicine. Very little was known about the structure of plants, and practically nothing of their manner of life. What was supposed was little enough and wholly wrong.

Following Aristotle, most botanists until the seventeenth century, accepted the idea that the life of plants derived from their possession of a " vegetative soul ". Aristotle had supposed that there were three kinds of " soul " or animating principle, the vegetative, concerned only with nutrition and growth, the animal, concerned with purposive movement, and the rational, which endowed its possessor with the power of rational thought. Plants neither moved nor cogitated, so they were possessed only of vegetative soul. This was connected with a further error, namely that the plant had no power to create its food, and it was supposed that it drew the material for its nourishment directly from the Earth in an already digested and preformed state. The Earth was regarded as the general " stomach " of plants, and the roots' only function was to draw up the nourishment the body of the plant required. Leaves were supposed to be protective organs, to keep the Sun and rain off the fruit, the production of which was supposed to be the proper end and meaning of the life of plants. Finally it was supposed, by some explicitly, that there was an analogy between plants and animals, though the extent of the analogy was a matter of dispute right through until the middle of the nineteenth century. Plants had their place in the general picture of a purposeful creation, existing for the nourishment, cure and pleasure of human kind. From the middle of the third century before Christ, until the early part of the fifteenth century, the literary tradition prevailed, and there is no evidence in the works that have come down to us that any of their compilers, with the exception of a very few people, early and late, compared what they read in the great classical herbals with what could be seen in the fields and gardens or in the apothecaries' shops.

The main classical sources were the works of two great botanists, both of whose works have come down to us in fairly complete form. The first was Theophrastus, the other Dioscorides. Most of the herbals that were compiled after them, and even many that were printed in fairly modern times, were essentially copies of their books, or compilations of fragments derived indirectly from them.

Theophrastus. He was born in Lesbos, probably somewhere about the year 370 B.C. He studied with Plato in the Lyceum, and after Plato's death became a pupil and close friend of Aristotle. It was to Theophrastus that Aristotle left his library and garden, and it was Theophrastus who succeeded Aristotle as head of the Academy. Aristotle's will has been preserved in the writings of Diogenes Laertius, and though Diogenes wrote much later than the time of these men, the will has a strong air of authenticity. It is known that Theophrastus took care of Aristotle's son after his father's death. Theophrastus wrote a great deal on a great many subjects, and a good deal of what he wrote has survived, though by no means all. His two great works on botany, the *Enquiry into Plants*, and the *Causes of Plants* have survived completely. The former has been translated by Sir Arthur Hort, and is published in the Loeb Classical Library. It remains an extraordinarily clear and interesting book to read.

Theophrastus is clearly battling with the actual real world of plants. He has assembled a vast amount of information, some he no doubt observed for himself, some Hort suggests would have been supplied by his pupils and assistants in the Academy, and some would have come from the scientific investigators which Alexander attached to the military expeditions of his conquest. The book describes the varieties, the locations, the structure and the uses of a very wide variety of vegetable forms, including many from remote regions. It is clear that Theophrastus was well aware of the differences between the flora of regions, and he makes constant reference to the habitat of the plants he is describing. His careful account of pines and firs, for example, not only distinguishes accurately between many species, but delineates the kind of situation in which these species and varieties grow, whether it is wet or dry, whether on highlands or lowlands, whether they favour a northerly or a southerly aspect, and so on. He even notices the chauvinism of some people who insist on calling their pines firs, because firs are generally regarded as having a pleasing aroma.

Behind the careful descriptions lies the ambitions of a notable classifier. Theophrastus wanted to group his plants into classes which would reflect their true similarities and differences, that is into classes which corresponded to the essential differences between plant forms. He struggles again and again with the plasticity of the material, with the difficulty of picking on some principles of classification which will work in a general way. At the very beginning of the book, in Book I, Chapter I, p. 3 he remarks : " In considering the distinctive characters of plants and their nature generally, one must take into account their parts, their qualities, the ways in which their life originates, and the course which it follows in each case : (conduct and activities we do not find in them, as we do in animals). Now the differences in the way in which their life originates, in their qualities and in their life-history are comparatively easy to observe and are simpler, while those shewn in their ' parts ' present more complexity. Indeed it has not even been satisfactorily determined what ought and what ought not to be called ' parts ', and some difficulty is involved in making the distinction". But not only was Theophrastus a notable classifier, he was also a genuine empirical scientist, and the cautious tone of this introductory passage is evident throughout the book. His classifications are rough, and he himself is well aware of it. He does not reach the essence of plant life and he himself draws our attention to this.

The three broadest classificatory principles that he employs, though very cautiously, are :

1. The identification of parts by analogy with the parts of animals :

2. The relative size and mode of life of plants, giving the fourfold division into trees, shrubs, under-shrubs and herbs :

3. The kind of locality in which particular plants flourish. Thus we might classify a tree as having such and such a kind of root, trunk, leaves, and liking a damp, northerly location.

The medical uses of plants form an important part of the Enquiry, though by no means all of it. It is in these matters that the strongly empirical character of Theophrastus's enquiry shows most forcibly. In medical matters magic has always loomed pretty large, and the gathering of herbs and roots for the preparation of remedies has always had a strong magical element in it. It was believed then, and indeed is still believed by many people today outside the scientific culture, that there are special ways in which herbs must be gathered in order that the gatherer should be protected and the herb effective in promoting a cure. Theophrastus expresses the usual cautious doubt of the true scientist, cautious because there might be something of value buried among the absurdities. He says in Book IX, Chapter VIII (p. 257, Loeb edition) : " Further we may add statements made by druggists and herb-diggers, which in some cases may be to the point, but in others contain exaggeration. Thus they enjoin that in cutting some roots one should stand to windward . . . These and similar remarks may well seem not off the point, for the properties of these plants are hurtful ; they take hold, it is said, like fire and burn ; for hellebore too soon makes the head heavy, and men cannot go digging it up for long ; wherefore they first eat garlic and take a draught of neat wine therewith. On the other hand the following ideas may be considered far-fetched and irrelevant ; for instance they say that the peony . . . should be dug up at night, for, if a man does it in the day-time and is observed by a woodpecker while he is gathering the fruit he risks the loss of his eyesight."

Copied, illustrated, broken up and reformulated, corrupted and rewritten this magnificent book became one of the sources of the herbal tradition.

Dioscorides : Pliny's *Natural History* was a potent source of plant lore for herbal compilers, but was itself a literary product, being a compilation from books rather than a description of nature taken from nature herself. Of greater importance and drawn in the first instance largely from nature was Dioscorides' *On Materia Medica*. This work was composed about 60 A.D. by a man who might well have been an army doctor. It was this book that was the most influential source for herbals, and was itself produced in innumerable editions, in many languages and in a wide variety of illustrated forms, from the beauty and accuracy of the Juliniana Anicia Codex written and illustrated before 500 A.D. to the absurd and stylized corrupt versions of the Latin Middle Ages.

On Materia Medica follows the pattern of Theophrastus in describing the plants and listing their habitats and medical uses. The illustrations formed an important part of this work and in the attempts by northern European painters to illustrate the editions in a way which could be used for the recognition of plants in their own regions, the revival of plant illustration began. Dioscorides' plants are mostly those of the eastern Mediterranean, and few of these plants flourish in northern Europe. To illustrate a herbal with pictures drawn from the English countryside and gardens, when the text concerns the plants of a quite different region, produced a vast confusion. Sometimes the herbal illustrator tried to find the plant that looked most like the one described sometimes that which had, locally, a similar traditional use. The result was a growing attempt to prepare herbals from nature, appropriate to the regions where they were to be used. This was in part a cause of the decline of the literary tradition. By the middle of the fifteenth century both in Germany and England, genuine native herbals with illustrations drawn from nature were being produced.

But with this revival of genuine natural study came a new development of magic, a development which was dominant, despite sceptical objections, for some 200 years, and which strongly influenced the practice of botany. This new magic was consonant with the rest of the

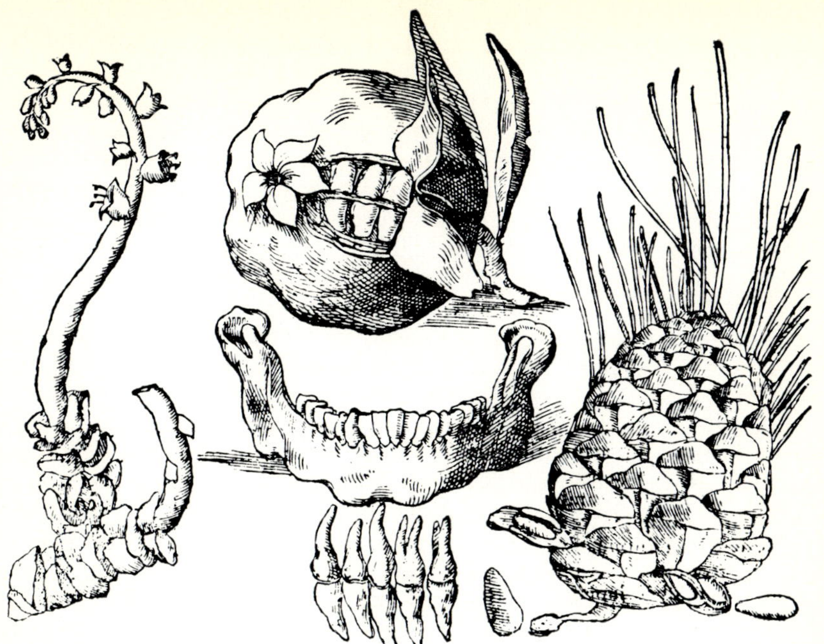

Fig. 8a. Plants resembling teeth.

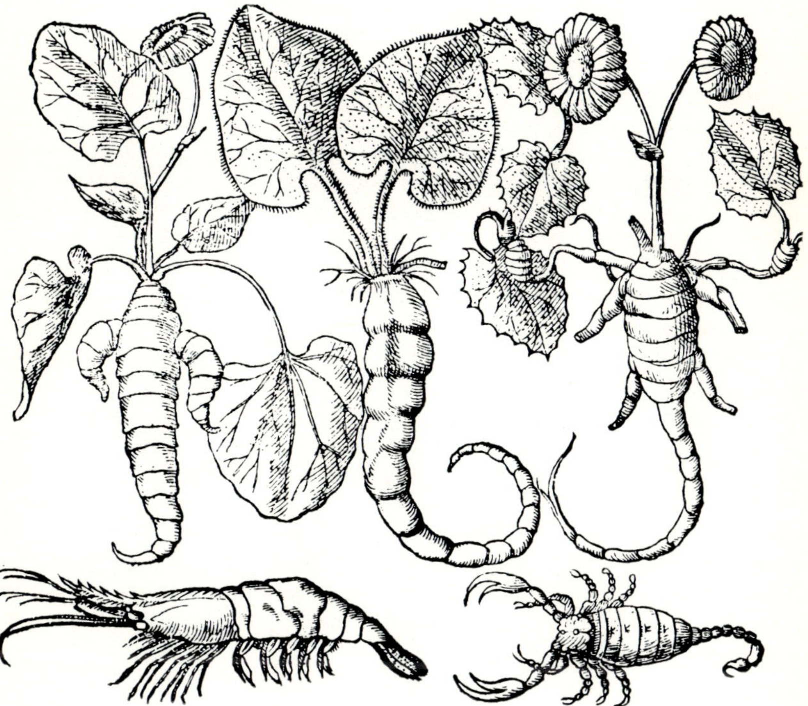

Fig. 8b. Plants resembling scorpions.

magical arts of the Renaissance. It was the doctrine of signatures, supplemented in ways we have learned about from the early history of magnetism, by astrology and Hermetism. The doctrine of signatures held that each plant that had medicinal or practical use had some resemblance in form to that for which it should be used. The invention of the doctrine is usually credited to Paracelsus, though there is little evidence that he orginated it. It is more likely that it was a traditional part of popular sympathetic magic.

Our old friend Della Porta was a convert to the doctrine, and wrote a very influential book promoting the theory. His *Phytognomonica* was published in Naples in 1588, and it contained a fairly detailed exposition of the theory. Reproduced in fig. 8a you can see his own diagram to illustrate the choice of plants that are good for the teeth, and in fig. 8b those that are good for relieving the sting of the scorpion. This peculiar theory was justified by such as Porta on the grounds that God put the signatures into the plants to guide us in finding cures. This He does in virtue of his general beneficence. The doctrine was propped up by two subsidiary principles of great importance. The signature or resemblance was a guide to what an extract of the plant was good *for*, thus in fig. 8a we see plants which resemble teeth, and hence are good *for* teeth. But it was also possible, by discerning the signature, to find out what the plant was good *against*. A scorpion-like plant, as illustrated in fig. 8b, was not a plant good for scorpions, but one good against the sting of that creature. God did not make the world for the benefit of scorpions, but for men, and hence the principle.

This doctrine was strongly maintained even into the seventeenth century, where it forms the basis of some, though significantly not all, of the work of a rather good botanist and influential herbalist, William Coles. He published two important books, *The Art of Simpling* (London, 1656), a short account of the kinds and uses of plants, and somewhat later a magnificent and botanically interesting book, *Adam in Eden*, a herbal in the true sense. In his *Art of Simpling* Coles offers a classification of plants that is no great advance on that of Theophrastus, nearly 2000 years before. On p. 14 he divides plants generally into the five categories. " 1. Trees, 2. Bushes, 3. Shrubs, 4. Herbs, 5. Neuters." Herbs are again subdivided, and on p. 15 we are offered : " 1. Potherbs, 2. Breadcorne, 3. Pulse, 4. Physical Herbs, 5. Flowers, 6. Grasse, and 7. those which in England we call weeds ". Clearly it is a classification by use and not by essential nature. Later, in Chapter XXVI he introduces a further principle of classification, by whether the plant is hot, cold, moist or dry.

In *Adam in Eden* Coles describes 342 plants. Each has a separate chapter, and is provided in most cases with a hand-drawn illustration, some of which are coloured. These are generally botanically accurate, as are his descriptions. The text for each plant is divided into the sections, The Names, The Kinds, The Form, The Place and Time,

The Temperature (and here he means the hotness, coldness, wetness or dryness) and the Signatures and Virtues in which the medicinal value of the plants is described. Now it is significant that by no means all the plants have a " signature " section, though all have one for " virtues ". For Coles, it is an empirical question whether or not a plant has a signature illustrating its virtue, and not a necessary condition for its having one. Coles is not a magician. It is the properties of the plant that endow it with virtue, not its resemblance to this or that organ. Nevertheless he thinks (p. 88) that " though Sin and Satan have plunged mankind into an Ocean of infirmities (for before the Fall, Man was not subjected to Diseases) yet the mercy of God which is over all his works, maketh Grass to grow upon the Mountains, and Herbs for the use of Men, and hath not only stamped upon them (as upon every Man) a distinct form, but also given them particular Signatures, whereby a Man may read, even in legible Characters, the use of them ". For instance, " *Heart Trefoyle* is so called, not only because the Leaf is Triangular like the Heart of a Man, but also because each Leaf contains the perfect Icon of an Heart, and that in its proper colour, *viz*. a flesh colour. It defendeth the Heart against the noxious vapour of the Spleen ". But Coles knew quite well that many useful plants did not have signatures. We must find out by practical tests, what are the virtues of plants " for Man was not brought into the world, to live like an idle Loiterer or Truant, but to exercise his mind in those things, which are therefore in some measure obscure and intricate." From these slowly developing empirically based herbals with their dawning apprehension of principles of classification a nomenclature evolved by which each plant received a pair of names, one for the general group to which it seemed to belong and the other which particularly seemed to describe it. Upon this foundation, at a later time, Linnaeus created the modern system of botanical classification and nomenclature. But into this, new knowledge was injected, coming from the anatomical and physiological study of plants.

The great work around which this part of the book is built is Stephen Hales' *Vegetable Staticks*. It is a study of the physiology of plants. It is to the growth and development of that study that we shall turn in the next chapter.

Further reading

Sir Arthur Hort's translation of Theophrastus is easily obtainable and makes interesting reading. For a more general account of herbals Dr. Arber's *Herbals* is the authoritative work. A short account can be found in A. C. Crombie, *Medieval and Early Modern Science* (Anchor A167 a, b), Vol. II, pp. 262–9.

CHAPTER 6

early modern ideas on the life of plants

WE have already seen how the development of the art of plant illustration led to a careful study of real plants and to more and more accurate reproduction of their main external features. Instead of copying the illustrations from older books the artists employed for the later Herbals drew from nature. The rapid growth of anatomical knowledge which this development brought forth was supplemented enormously when in the seventeenth century the microscope was brought to the study of plant structure. Robert Hooke observed that plants were cellular in structure, and it is to him that we owe this use of the word " cell ". It became clear that the plant had a complex internal structure, with cells not only arranged in various organs, but also differentiated amongst themselves. The most industrious and accurate plant anatomist of the time was Nehemiah Grew. In fig. 9 you can see a reproduction of his drawing of the cross section of a branch of pine. This is taken from his remarkable book *The Anatomy of Plants*, which was published by the Royal Society in 1682. In Grew's many studies the vascular system of plants became clear. There are vessels within a plant along which both air and watery fluids can move. How such movement is initiated and how it contributes to the life of the plant became one of the chief problems for plant physiologists for 200 years.

The students of plant life were faced with two main sets of problems. How did plants reproduce themselves ? How did they feed and grow ? Aristotle knew that many plants reproduced themselves, though many people continued to believe, until relatively modern times, that some plants were spontaneously generated. Many plants, it had been observed, reproduced by budding. Was the production of seeds a rather special way of budding, or did it have something in common with the way animals reproduced ? If the seeds were part of a process of reproduction essentially similar to that of animals, were plants differentiated into male and female ?

There were several factors tending to produce confusion in the attempts to answer these questions. Animal reproduction was not clearly understood. It was widely believed that the role of the male was the production of seed which germinated in the female. Thus the plants were seen as male, and the earth as the female and the mother of all plants. That there could be female *plants*, was thus an inconceivable idea. The problem was further confused by the tendency, in popular plant lore, to speak of strong, robust plants as male and for generally

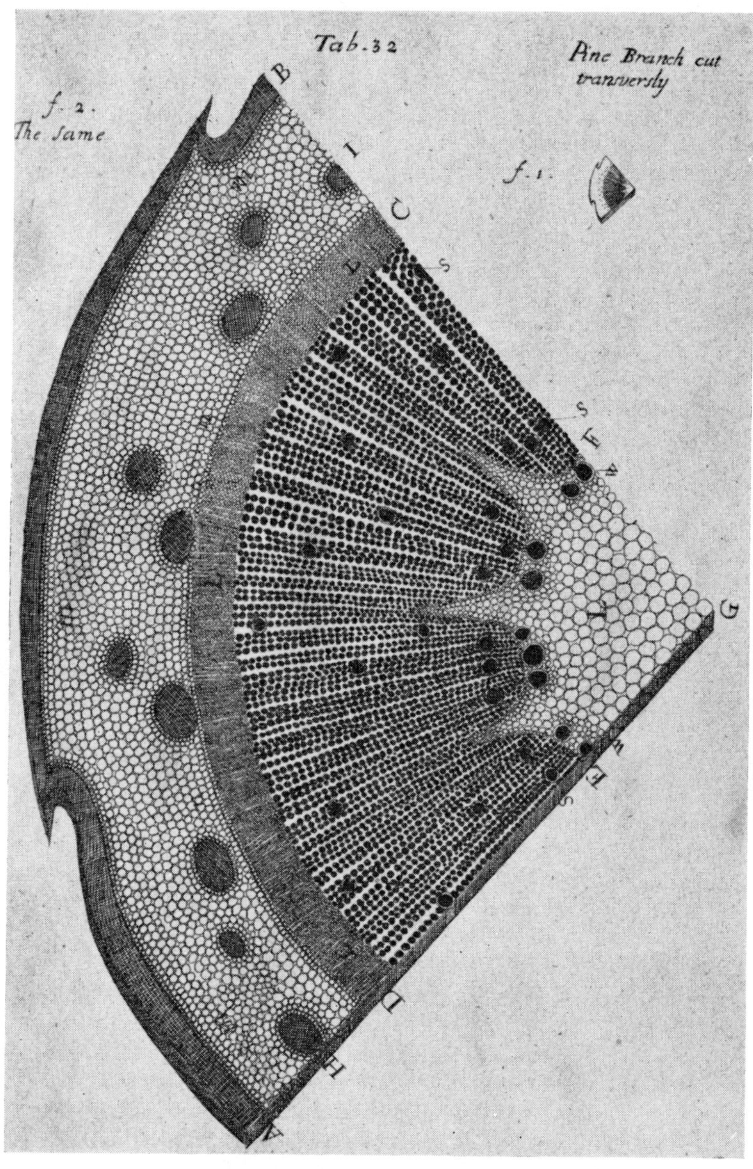

Fig. 9. Pine branch cut transversely.

softer and more delicate plants as female. Several ancient botanists differentiate species in this way; for example, among pine trees distinctions in external form and in general appearance were marked by calling the more "spiky", male, and the more "feathery", female. However, Theophrastus had noticed that some branches are fruit-bearing and others not, and other botanists had noticed that the same was true of different individual trees within an orchard. But these observations remained unconnected with any theory. Even in the sixteenth century, Cesalpino, the last of the distinguished Aristotelian botanists, was quite ignorant of the role of the pollen in plant reproduction. He took the traditional view that the propagation of seeds was a nobler form of budding. A seed was essentially a detachable bud. Having no idea of the part played by pollen, he had no idea that seeds were embryo plants themselves, the product of a complicated, essentially sexual mode of reproduction.

The hypothesis of the sexuality of plants and the true view of their mode of reproduction we owe to Sir Thomas Millington. It was Millington who first identified the stamens as the male organs of generation in plants. In this he was quickly seconded by Nehemiah Grew. Speaking of pollen, Grew says on p. 171 of *The Anatomy of Plants*, . . . "the *Globulets* or small *Particles* within the *Thecae* of the *Seed-like Attire*, and upon the *Blades* of the *Florid*, I have conjectured that they are the *Body* which *Bees* gather and carry upon their thighs . . ." Then one day, "In discourse with our Learned *Savilian* Professor Sir *Thomas Millington*, he told me, he conceived that the *Attire* doth serve as the *Male*, for the *Generation* of the *Seed*. . . ." Neither of these men had the details quite right. Grew goes on: "And the *Globulets* and other small *Particles* upon the *Blade* or *Penis* and in the *Thecae*, are as the *Vegetable Sperm*. Which, so soon as the *Penis* is exerted, or as the *Testicles* come to break, falls down upon the *Seed-case* or *Womb*, and so touches it with a *Prolific Virtue*". It is clear that Grew, and perhaps Millington too, thought that the pollen fertilized the female cells of the plant within which it was produced. It seems not to have occurred to him that the bees, which he has just commented upon, carry the pollen to other flowers. The plant is a hermaphrodite, but it does not fertilize itself.

The final proof of the role of the pollen and the stamens and anthers seems to have been provided by Camerarius, who in 1691 to 1694 showed that no seed capable of germination could be produced without the co-operation of pollen. He is quoted by J. Von Sachs in his *History of Botany* p. 387 as writing: "When I removed the male flowers of *Ricinus* before the anthers had expanded, and prevented the growth of the younger ones, but preserved the ovaries that were already formed, I never obtained perfect seeds, but observed empty vessels, which fell finally to the ground exhausted and dried up. In like manner I carefully cut off the stigmas of *Mais* that were already dependent, in consequence of which the case remained entirely without seeds, though the number of abortive husks was very great ".

This discovery was not immediately followed up, and was indeed vigorously denied by many botanists. It was not until the nineteenth century that the sexuality of plants was finally accepted as an established fact by all serious botanists, and the details of the process by which the pollen grains penetrate the style, and the way in which gametes are brought together in the formation of the seed discovered.

The problems of the nutrition of plants provided a more complicated set of questions than the problem of the reproduction. There were really three interlocking questions, correct answers to all three of which were required before the basic theory of plant nutrition could be correctly formulated.

1. Did plants draw their nutriment fully formed from the Earth, or did the plant itself elaborate simple elements into the substances, known to be diverse, of which a mature plant consisted ? Aristotle, and certainly his followers, right down to the seventeenth century, believed that the plant drew fully prepared substances from the Earth.

2. What were the functions of the various organs of the plant ? What were leaves for ? Did the roots absorb only water ? If not, how did they distinguish good from harmful substances in the soil ? Was this the function of their vegetative souls ? It was widely believed that the leaves were to provide shelter and shade for the developing fruit.

3. Did the sap circulate within the plant ? If it did, what was the power that moved it ? And did it have a pulsating circulation like the blood of animals ?

The work of Stephen Hales, which we shall study in some detail and which forms the centre of this part of our study, provided the correct and final answers to several aspects of these questions, though, as with Grew and Millington, the final acceptance of his results was rather long delayed. But Hales was the Newton of plant physiology, and like Newton, stood on the shoulders of giants.

Though he was the last of the Aristotelians, Cesalpino was also the first modern plant physiologist, and in his remarks on the nutrition of plants he moves slightly but perceptibly away from the Aristotelian theory. There is a strong echo of Aristotle in Cesaplino's view that plants are not hot because their food goes entirely to body building, and they use none for actively running around or for ratiocinative thought. They have no need to think, since they do not have to find food, nor do they have to be able to recognize what is food and what is not, so they have no organs of sense. Cesalpino also makes much of the growing analogy thought to obtain between plants and animals, and he declares that sap is like blood, oozing along narrow vessels in the stem and roots of plants. He cites the bleeding of figs and of vines as showing that there is a movement of sap within plants. There is no hint in his ideas yet of a circulation of sap. It seems clear that that idea had to wait for

Harvey's prior demonstration of the circulation of blood in animals. Cesalpino is yet further from Aristotelian ideas in his finding it puzzling how the roots can draw watery sap from the Earth. He rejects the view that it is similarity of nature that draws water into the watery plant. Nor does he find the impulsion to fill a vacuum adequate. His own explanation is to draw an analogy between the woody texture of roots and the nature of sponge. It is the essential dryness of the roots which, sponge-like, draws water into them from the surrounding damp soil. All this is crude and speculative, and based upon a little, a very little common observation.

The first true experiment in plant physiology was performed by J. B. Van Helmont in the early part of the seventeenth century, an experiment that still finds a place in botanical textbooks. Van Helmont weighed a young willow tree, and planted it in a weighed pot of earth. After five years he weighed the tree again, and reweighed the earth. The tree had increased in weight by some 100 lb., while the earth had decreased in weight by some two ounces. The material of the tree could not have come from the material of the earth. Van Helmont concluded that the bulk of the tree must have come from the water with which the pot had been regularly irrigated, and that it was by the elaboration of water that the tree had created its own peculiar materials. Boyle performed a further series of experiments designed to conclusively disprove the Aristotelian theory of preformed material in the earth.

In Boyle's experiments the hypothesis under attack is the idea that the kind of materials in the plant derive their qualities preformed from the earth. Van Helmont simply showed that the *bulk* of the plant material came from the water, but it remained possible that in that two ounces drawn from the earth all the trees qualities were contained, like the oiliness of the olive, the sweetness of the ripe pear, the whiteness of the blossom and so on. Boyle performed two main experiments in refutation of that theory. In the first he observed the grafting of pears upon whitethorn stocks. The fruit of the whitethorn is a sourish berry, whereas the product of the pear graft is a sweet and agreeable fruit. Since the whitethorn root system is common to both productions it is hard to see how the root system could select sour qualities for the white-thorn berry and sweet qualities for the pear. The sourness and the sweetness must derive from the different elaborations of similar basic materials within the plant in the process of formation of the fruit. The stuff of plants is not drawn, fully formed as to qualities, from the earth. In the other experiment Boyle grew plants in water alone, without any earth. He found that he could distill an oil out of the plants so grown which would not dissolve in the water in which they were grown. This oil had different qualities from the water. Boyle concluded that these new qualities had been produced as the result of elaborations of the elementary particles of water within the plant and not been drawn pre-formed from water. They were not qualities of water. Thus between them Van Helmont and Boyle had shown that neither the bulk nor the qualities of the material of plants were drawn from the earth, but it seemed that water played an essential role.

The roots draw water from the earth, and somewhere in the plant this material, and perhaps other materials, are elaborated into the stuff of plants. All this argues for a circulation of sap within the plants, upwards to the leaves, and perhaps downwards again to other parts. The important idea that there is a circulation *downwards*, from the leaves to other parts of the plant, as well as a movement *upwards* from the roots we probably owe to J. D. Major of Kiel, who, following Harvey, proposed in 1665 that the sap of plants too circulates. It was not long afterwards that Malpighi in 1671 realized the true function of the leaves, that it is in the leaves that the elaboration of the simpler elements like water into the complex stuff of plants took place. This being so there *must* be a return movement of sap from the leaves to the rest of the plant. This, in Malpighi's view, did not have the regular rhythmic character of the circulation of the blood, but was a more uneven process both slower and subject to seasonal change. Not only did Malpighi realize the true function of the leaves but he was also aware that the material elaborated in the leaves is stored by the plant for future use, so that there are subsidiary movements of sap, in which stored material is recirculated to serve the purposes of later growth.

The correct idea of the movement of sap, went along with and was inseparable from the correct idea of the function of leaves. However, the only experiment that Malpighi seems to have performed on the hypothesis of the role of the leaves depended on a further bold hypothesis. He argued that the cotyledons of young plants are true leaves, and he showed that the young embryo plant will not develop if the cotyledons are detached. He argues by analogy that the leaves of the adult plant also are essential to its growth. However, he was not clear as to the connection between green colour and food elaboration in plants, and had no idea at all of the role of chlorophyll. This led him to suppose that not only the leaves but the stem too had a food-elaborating role. But this, as we know, is true only of those plants which have green stems, and hence chlorophyll cells in the stem surface.

It had been realized that some of the vessels within plants contained air. But it had not been realized that the atmosphere had an essential role in plant nutrition. It was Malpighi who first understood that plants breathe, though the chemical knowledge of the time was insufficient to enable him to distinguish between the inhalation of carbon dioxide and the exhalation of oxygen as a food-absorbing process, and the inhalation of oxygen and the exhalation of carbon dioxide as a true respiratory process, both processes being essential to the life of plants. But Malpighi's friend, Nehemiah Grew, realized that the atmosphere must contribute *food* material to the life of the plant. The state of knowledge of plant nutrition before the great work of Stephen Hales can be gathered in summary from a selection of passages from Grew's *Anatomy of Plants*.

1. Argument against the pre-existence of plant substance in the earth. " So the *Principles* of the Purgative Parts of a *Root*, as of Rhubarb, although we should suppose them to be pre-existent in the surrounding *Earth*, yet we cannot say, that that *Earth*, or the Principles contained therein are Purgative ; but only that they are such, as by being combined together, in such a peculiar way, may become so. So the several parts of a *Clock*, although they are and must be all pre-existent to it, and it is their *Form*, by which they are what they are ; yet it is the *setting together* of such *Parts*, and in such a way only, that makes them a *Clock*."

2. Grew's realization that the air contributes to the food of plants is contained in this passage : " *Earth*, *Water*, *Air* and *Sun* ; all which, in that they contribute so universally to *Vegetation*, and to whatsoever is contained in a *Vegetable* . . ." are essential. But various facts " seem to argue that the air is impregnated with *Vegetable Principles* ", that is with something which is used in the elaboration of plant material.

3. Finally Grew applied the Corpuscularian ideas of his time to the understanding of the structure of plants. The Corpuscularians had argued that all the differences between different things and different substances could be put down to differing arrangements of tiny, microscopic corpuscles or particles. Thus a thing had a smooth surface if the particles were arranged in even, level lines, but would be rough if they were arranged in humps and peaks. Grew tries to explain the structures that he sees, either with the naked eye or with his microscope, as the arrangement of small particles of the chemicals, salts, which go to make up the substance of the plant. Talking of the arrangement of the molecules, as we should call them nowadays, in the development of the leaf vessels, he says : " For if the *major part* be applied End to End, and only every Third or Fourth applied End to Side, they produce a great *Circle* . . . and if the Application be the same, but to the contrary Side, they thence begin a new *Circle* with the same *Diameter*, but with another *Centre*, answerable to the intended *Shape of the Leaf* ". *The Anatomy of Plants*, p. 160, Section 17.

Further reading

There are disappointingly few histories of botany and of the life of plants. The only book which can really be recommended is that of J. Von Sachs, *History of Botany* translated by H. E. F. Garnsey and I. B. Balfour.

Practical Work

1. The classical experiment of Camerarius can easily be repeated using the common campion (*Lychens*), which has male, female and hemaphrodite flowers. The female flowers should also be protected from outside pollination by isolating the whole plant.

2. Using a modern botanical text book and *working from* some suitable flower, such as the flowers of the daffodil or iris, work out what are the modern terms for Grew's names for the male and female parts of a flower.

3. Weigh some earth, put it in a small pot, and plant in it a bean which has already been weighed. Weigh bean and bean plant, and the earth again when the bean has germinated and produced a plant with two or more green leaves. Does your result bear out Van Helmont's experimental observation on the willow ?

4. Compare the result of placing beans on wet cotton wool in the case where the bean is not broken up, and the case where the embryo plant is detached from the cotyledons. You might also try an experiment that Malpighi did not do, namely detaching the green leaves as they develop from one of the bean plants you used in the third experiment, and see what happens.

5. Detach some quantity of a plant which has been grown only in water such as water lily. Using a very gentle heat, first drive off the water in the plant, and then collect some of the oily, tarry substance that comes off in further gentle heating. Will this mix with water ?

CHAPTER 7

the life of Stephen Hales

Green Teddington's serene retreat
For philosophic studies meet,
Where the good Pastor Stephen Hales
Weighed moisture in a pair of scales,
To lingering death put Mares and Dogs,
And stripped the Skins from living Frogs.
Nature, he loved, her Works intent
To search or sometimes to torment.

From *The Boat* by Thomas Twining.

STEPHEN HALES was born at Bekesbourne, Kent, on 17 September 1677 of a well-to-do, upper middle class family, somewhat similar in circumstances and history to the family of William Gilbert. Little is known of his boyhood. In 1696 he entered Bene't College, Cambridge. This was the college that later came to be called Corpus Christi. It was not a large college by Cambridge standards, having only 12 fellows. However, the interests of the fellows were diverse, and the undergraduates were able to study a remarkably wide range of subjects, both literary and scientific. These were taught in an up-to-date fashion. They read Locke's *Essay*, for instance, as their text in logic. The college had a particular interest in the bio-medical sciences, and was very ready to encourage research, even amongst junior members, though without the means or the grandiose ambitions of Bentley's Trinity College, with its observatory and its chemical laboratory. Hales showed considerable academic promise and was pre-elected into a fellowship in 1699. This meant that he would fill the next vacancy amongst the fellows. At that time his studies were mostly theological. A vacancy did occur in 1703 and on 25 February he was formally elected to a fellowship. Following the usual steps in an academic career in those days he was ordained on 19 June 1709.

In the meantime however a most remarkable young man had come up to Bene't College, a man whose influence on Hales was lifelong. This was William Stukeley who came up in the autumn of 1703. He had an intense interest in botany, in anatomy and in physics and chemistry. In his early days at Cambridge he tells us that he " went frequently a sampling [i.e. collecting herbs], and began to steal dogs and dissect them and all sorts of animals that came our way. We saw too many Philosophical Experiments in Pneumatic Hydrostatic Engines and instruments

. . ." He studied "the doctrine of Optics and Telescopes and Microscopes and some chemical experiments, with Mr Stephen Hales, then Fellow of the College". Hales and Stukeley very quickly became friends. "They rambled," says a contemporary, "over Gogmagog Hills and the bogs of Cherry-Hunt Moor to get samples. . . . They proceeded also to the dissection of dogs. . . . They applied themselves also to chemistry. . . ." Hales often provided the ideas for experiments which Stukeley carried out. One of their most remarkable was the use of molten lead to fill a pair of lungs. When the lead had solidified they allowed the lungs to rot away, revealing a lead copy of the internal structure. Stukeley rather ruefully describes that he broke it up bit by bit giving pieces away to friends. Stukeley left Cambridge in 1709, and this extraordinary activity came to an end. There must surely be some connection between the fact of his going and Hales' own decision to leave Cambridge and take a small country parish. Shortly after his ordination Hales became the Perpetual Curate of Teddington, being appointed to the parish on 10 August 1709. Here he pursued his scientific work in the intervals of his priestly duties, for the rest of his life. He steadfastly refused all offers of advancement in the Church so as to be free to pursue his scientific work. This began almost immediately he arrived at Teddington.

Though we shall be studying his work on the physiology of plants he was equally devoted to studies on the physiology of animals. In both fields his method was similar, that is to apply the methods of physics and chemistry to the study of living creatures. He believed that God had ordained the laws of physics and chemistry in accordance with which everything in the Universe worked, and so his own special creations of plants and animals could also be expected to obey those laws. It was in this spirit that he began his first major work on the circulation of the blood. Harvey had shown in rough outline that the blood must be supposed to circulate, but very little was known about the details of the process. Certainly not enough was known to decide whether the contraction and swelling of muscles was the effect of blood pressure. It was really in pursuit of the answer to that question that Hales did his very detailed studies on *Haemastatics*.

The first problem was to form some fairly accurate idea of the blood pressure in a variety of animals. By some carefully performed insertions of tubes in the blood vessels of various animals Hales succeeded in making comparative measurements for a good many animals. He discovered some important generalizations which remain an accepted part of comparative physiology. The pulse is more rapid with small animals, but the blood pressure is roughly proportional to the size of the animal. It follows that the pressure is not a function of the rate of the pulse. It has something to do with the amount of blood expelled at each beat. To find this quantity Hales injected sufficient soft wax into an empty but fresh heart to fill the ventricle and to open valves. This he reasoned would be roughly the amount of blood that was present in the ventricle when the valves opened. From the amount of wax he was able to estimate the quantity of blood expelled at each beat. He found that the output of the heart was relatively greater for the smaller animals.

He also investigated in detail the circulations and quantities of blood moved in the various sub-systems of the circulation. This study led him to one of his most beautiful investigations. By a careful study and measurement of the blood vessels, including the finest capillaries, and careful quantitative studies of the surface of the alveoli in the lungs, he was able to predict, by calculation, that the rate of flow of the blood would be very considerably greater in the lungs than in other parts of an animal. He confirmed this by microscopic studies of the lungs and muscle capillaries of live frogs, and he found, by measuring the rate at which a single blood corpuscle passed along a capillary, that the rate of blood flow in the capillaries of the lungs was five times the rate in the muscle capillaries, a confirmation within the appropriate order of magnitude of his theoretical calculation. The results of all these experiments he published some 20 years later in his *Haemastatics*.

It is evident from the parish records that Hales brought the same methodical efficiency to the affairs of his parishioners as he did to his science. He seems to have been an exemplary clergyman. The records are kept in his own hand, and there are two important parish works of his time. In 1716 he organized the enlargement of the church and the repair of its roof and paid for these improvements by selling pews. Since many pews were already owned by parishioners the adjustment and organization of the seating of his parish church presented a problem hardly less testing than the problem of the circulation of the blood in the lungs. In 1754 he undertook to improve the village water supply and by a steadily inceasing exploitation of the local springs he succeeded in producing a flow of 60 tons per day, the quantity determined in a characteristically Halesian way.

It was the custom in those days for the more senior or distinguished clergy to hold more than one living, and to employ a deputy to look after the parishes in which they did not reside. It was, however, not permitted to hold a plurality of livings and a college fellowship, so when Hales was given the living of Porlock in Somersetshire, in addition to his parish at Teddington, he was obliged to resign his fellowship. It seems that this brought his connection with Cambridge to a complete end, since he does not seem to have visited the city ever again. However, in 1722 he resigned the living at Porlock in favour of appointment to Farringdon near Winchester. He did not treat this parish solely as a source of income, and visited it every year, staying for a couple of months or so. Perhaps this was because it was near Selborne, where the White family lived. Hales was indeed friendly with the Whites and particularly Gilbert White the celebrated author of the *Natural History of Selborne*. Apart from his biophysical interests Hales does seem to have been interested in all sorts of problems related to biology and to its practical application in agriculture. His interest extended to the application of machinery to agricultural tasks and he invented a successful winnowing machine.

In the same year that he resigned his fellowship Hales was elected a Fellow of the Royal Society, and it is clear that by this time he had acquired a considerable reputation as a physiologist, both of animals and plants. This reputation was not only amongst other scientists, as the poem at the head of this chapter witnesses. Without anaesthetics his experiments on the blood pressure and the circulation of animals were, by our standards, unbelievably gruesome and cruel. Mares were tied to gates and pipes inserted in their jugular veins, frogs were skinned alive so that the contractions of their muscles could be observed, and so on. In the context of the times these were not particularly notable cruelties, though more sensitive people, such as Pope, were beginning to find contemporary cruelty to animals obnoxious. There must have been something in Hales' general style of life and in his biological interests that disarmed the opponents of vivisection, since Hales became quite a close friend of Pope, after they became neighbours. He advised Pope on horticultural matters, and was one of the witnesses of his will. Pope somewhere mentions his distaste for the way Hales had to do his experiments but nowhere suggests that they should be stopped. And, of course, what inestimable benefit to mankind has derived from Hales' discoveries of the true facts of circulation !

In March of 1720 Hales married Mary Newce. She died on 10 October 1721. After this sad event his devotion to the simple duties of the parish and his systematic exploration of the problems of plant and animal physiology became his whole life for a while, though, as we shall see, other interests of a practical nature from time to time engaged his attention.

John Mayow had discovered oxygen about 1670. He called it " spiritus nitro-aereus ", and in the course of his experiments came to realize that respiration and combustion were forms of the same process. Hales knew of these experiments and turned his attention too to the role of the air in life. In Chapter 9 we shall look in detail at his experiments. He repeated Mayow's work and though he misinterpreted it he made the further discovery of carbon dioxide. Experimenting on this substance he found ways of absorbing it and so of purifying the air. This led him to the invention of the respirator, of which he devised several kinds. As we shall see, the purification of the air became one of his major preoccupations.

In 1727 he was elected to the Council of the Royal Society and from that time on served on a great variety of bodies with public interests connected with the Society or its members. It was this which led him into the study of the " distemper of the stone ". This was a painful condition in which small hard bodies, called renal calculi, were formed in the bladder and urinary tracts, fortunately a rare condition nowadays. He had himself attempted without success to find a solvent for these bodies which could be taken safely by a patient. Failing in this he invented a surgical instrument for the physical removal of the stones. This instrument became widely used. In 1733 he drew together the results of his series of experiments on the blood circulation, and published them under the title *Haemastatics*.

From then on the practical application of science became his chief interest, though he was involved in other public affairs as well. In 1732 he became one of the Trustees of the Colony of Georgia, and was very actively associated with the development and management of its affairs. The venture was one of mixed philanthropic and strategic interest. The Trustees, such as Hales, had thought of the colony as a place to which worthy debtors and their families could be sent to start a new life, and also as a place for refuge for persecuted Protestants from the Continent. They collected fairly large sums of money for the transport and establishment of the people chosen. At the same time the Government had an interest in occupying the region with settlers to act as a buffer against any possible Spanish encroachments from the south. These mixed reasons for establishing the State of Georgia led to the growth of two factions amongst the trustees, the philanthropic and religious group, of which Hales was a prominent member, and the political and commercial set who opposed the use of any money for the establishment of religion or of libraries. On the whole the philanthropists had the better of it though the battles were often bitter.

The gin problem next engaged Hales' attention. And here he was able to employ his talent for the scientific backing up of publicly desirable policies. Cheap gin had become a menace of frightening proportions in England, and Hales and many others were concerned to bring about a taxation bill which would cut the consumption of this spirit by making it prohibitively expensive. A characteristic Halesian touch about his arguments was provided by his calculation that the landed interests were losing £6 000 000 annually through the decline in food consumption among gin drinkers, Hales wrote two tracts against gin, one in 1734 and another in 1736. The efforts of Hales and others in fact bore fruit and the Gin Act of 1736 did have considerable effect in reducing this menace.

His early interest in the problem of renal calculi led to his taking a part in examinations of alleged cures. In 1739 he was one of a large commission empowered to investigate the claims of Joanna Stephens to have found a medicine for this condition. Hales was not enthusiastic but his experiments showed that there might be something in it. The commission did in fact award Joanna Stephens £5000. The *Liquid Shell* remedy was not so fortunate, being exposed by Hales as useless. In the same year as his examination of the Stephen's medicine he was awarded the Copley Medal of the Royal Society.

The Georgia affairs in which he was interested depended very much upon shipping, and Hales began to investigate the problems of life at sea, particularly the problems raised by the supply of pure food, water and air. In 1740 he published *Philosophical Experiments*, a series of investigations into these problems. Amongst other devices, in 1739 he invented a way of measuring the depth of the sea where the water was too deep for the use of a line. The device consisted of a copper ball and a tube inserted into it, with its outer end open. A plug of fatty material was inserted in the tube, and the ball, attached to a weight, dropped into the sea. When it reached the bottom the weight was automatically released and the ball rose to the surface. The fatty plug was pushed up the tube by the water pressure, by an amount proportional to the depth, and though it was pushed out again as the air in the ball expanded with the release of pressure as it rose, there was a ring of oil deposited in the tube showing where the plug had been. The device was tested in places where a line could also be used and shown to be accurate. Unfortunately when it was tried at very great depths the weight did not release and the whole thing was lost.

Hales' experiments on the air had made him very aware of the importance of ventilation to good living conditions. From 1740 till the end of his life he worked upon the problem. He began with the problem of the ventilation of ships, having been sent to help with an outbreak of sickness aboard Lord Cathcart's fleet as it waited at Spithead to set sail for America. It was clear to Hales that part of the trouble was the foul air in the living quarters of the ships. At that time most people believed that many diseases were caused by poisoning with bad air. Hales proposed the use of large bellows which would extract the air, and pure air would then flow down into the ship through appropriate ports. The idea had been proposed practically simultaneously by the Swedish naval architect Triewald, who applied it immediately to the ships of the Swedish navy. Hales' ventilators were not applied so immediately to the British navy because a rival system had been invented by Samuel Sutton, using a much simpler means of moving the air. Sutton's idea was to insert long tubes down into the ship and to force an upward draught of air by heat, using the ship's own galley fires where possible. These tubes in fact worked quite well, though they did occasion some anxiety by those who used them because of the fire danger, always a problem on wooden ships. For a while Sutton had the best of it, since his apparatus was a good deal cheaper to install. To promote the bellows idea Hales wrote his *Description of Ventilators* in 1743. In the end, despite considerable efforts by Hales and some influential friends the ventilators did not become generally used by the Navy. However, they were very widely used, both in the Merchant Navy and in the slave trade, where their use greatly ameliorated conditions on long oceanic voyages. There still remained a problem, how to see to it that the sailors or slaves pumping the bellows kept at their work when they could not see the air issuing from the vents. To solve this psychological problem Hales installed windmills in the apertures of the vents, and when these were invisible they had little bells fitted to them so that the pumpers could hear definite evidence of the efficiency of their efforts. But even these were not enough, so he suggested and built windmill powered ventilators, which, provided they were adequately maintained, worked remarkably well. This innovation was made in 1751. All these practical advances were summed up in his *Treatise on Ventilators* of 1758. Not only were the ventilators used on ships but they soon found a use in hospitals, prisons and mines, where the problem of bad air was at least as bad as on ships. The windmill of his ventilator on the Fleet Prison remained one of the sights of London for many years. Clean air went with clean rooms in Hales' mind, so in fact under his influence a great deal more sanitary conditions obtained in the prisons and hospitals of England for a while. Yet squalor returned, and much as Mayow's discovery of oxygen and much of Hale's own scientific work were temporarily lost, so were Hales' great sanitary reforms.

In his old age Hales no longer lived alone at Teddington but had as his housekeeper his niece Sarah Margaretta Hales. Ventilation did not wholly absorb his energies. In 1743 he undertook a major rebuilding of Teddington Church, and the old church is still much as he left it. In 1747 he made extensive tests on the efficacy of alleged wonder medicines, particularly the famous Tar Water, and another supposed Purging Water. He was not enthusiastic. His interest in practical applications of science led him to serve as one of the founders of the Royal Society of Arts in 1754. In 1756 he made yet another practical invention of revolutionary importance, the forced distillation principle. He found that blowing a current of air through a boiling liquid would greatly speed the distillation process, saving fuel and time, and so cost. His last experiment was on the way fish breathed, and was done in1756.

In his later years he became friendly with Prince Frederick, who used to ride over to visit him ; Princess Augusta also knew him well. He became her chaplain. But throughout all these years, as a Trustee of Georgia, a member of the Council of the Royal Society, a prominent reformer of the conditions in the hospitals and prisons of England he remained Vicar of Teddington, entering the births deaths and marriages of the people of his parish with his own hand. He died on 4 January 1761.

Further reading

There is a short account of his life in the introduction to the Oldbourne edition of *Vegetable Staticks*. His *Life* by A. E. Clark-Kennedy is a really fascinating scientific biography, and gives a very detailed picture of the times in which Hales lived. Called *Stephen Hales*, this excellent book was published by Cambridge University Press in 1929.

CHAPTER 8

the sap

THE idea that water and other materials came into the plant through the roots and passed upwards to its other parts was a commonplace from antiquity. Indeed, it had been supposed that that was the only vital process in the life of plants. No-one, however, had attempted to discover the dynamics of this process, that is, to relate the amounts absorbed to the amounts perspired (or transpired, as we should now say), and to relate both of the quantities to time. Van Helmont's experiment was strictly static and had to do only with the material make-up of a plant. However, both Malpighi and Grew had formed the idea that there was both movement of fluid from the roots to the rest of the plant, *and* from the leaves to other parts. Grew's careful anatomical studies had revealed the certain existence of vessels within the plant, capable of the transport of both watery substances and aeriform ones. Thus the hypothesis of *movement* and flux and reflux of sap was formulated. But Harvey's hypothesis, that the blood of animals circulated, was the basis, by analogy for a *circulation* picture, for the movement of the sap of plants, often seen as analogous to the blood of animals. For various reasons it was supposed that the transport of sap was upwards in the inner part of the stem, and downwards in the other part. Perhaps what really lay behind this hypothesis was the idea that in the young plant the transport must be from the root to the plant proper, and so since the inner part of the stem represented the oldest part of the plant, the embryo direction of sap movement would persist. Whatever the reasons, it was widely held that sap moved upwards in the inner part of the stem and downwards in the outer. But *why* was there any movement of sap at all ? How was this sap movement " powered "? The Aristotelians, of course, thought of the sap as essentially watery, and so inclined to find its natural place above the heavy earth, so watery sap would of its own nature flow upwards. It was into this complex of problems that Hales first brought systematic experimental exploration, dealing in turn with the total water economy of the plant, the pressure of the sap within it under various conditions, and the movement or circulation of the watery substance.

The first thing to find out is the *amount* and *rate* of water movement, considered under the three heads of *root*, *stem* and *leaves*. Hales did not take it for granted that the water entered the plant at the root and left it at the leaves, but in order to make a beginning with quantitative experiments he began with that assumption. Supposing that that was the general direction of water movement, he explored the process quantitatively in a series of experiments. We can get a good idea of the style of these experiments by looking at Experiment II of Chapter 1.

" From *July 3rd* to *Aug. 3d.*", he says, " I weighed for nine several mornings and evenings a middle sized *Cabbage Plant*, which grew in a garden pot, and was prepared with a leaden cover as the Sunflower, *Exper. 1st.* Its greatest perspiration in 12 hours day was 1 pound 9 ounces ; its middle perspiration 1 pound 3 ounces, = 32 cubic inches. Its surface 2736 square inches, or 19 square feet. When dividing the 32 cubic inches by 2736 square inches, it will be found that a little more than the 1/86 of an inch depth perspires off its surface in 12 hours day."

" The area of the middle of the Cabbage stem is 100/156 of a square inch ; hence the velocity of the sap in the stem, is to the velocity of the perspiring sap, on the surface of the leaves, as 2736 : 100/156 : : 4268 : 1 for 2736 × 156/100 = 4268. But if an allowance is to be made for the solid parts of the stem (by which the passage is narrowed) the velocity will be proportionably increased."

" The length of all its roots 470 feet, their periphery at a medium 1/22 of an inch, hence their area will be 256 square inches nearly ; which being so small, in proportion to the area of the leaves, the sap must go with near eleven times the velocity through the surface of the roots, that it does through the surface of the leaves."

"And setting the roots at a medium at 12 inches long, they must occupy a hemisphere of earth two feet diameter, that is 2.1 cubic feet of earth."

" By comparing the surface of the roots of plants, with the surface of the same plant above ground, we see the necessity of cutting off many branches, from a transplanted tree : for if 256 square inches of root in surface was necessary to maintain this Cabbage in a healthy natural state: suppose upon digging it up, in order to transplant, half the roots be cut off (which is the case of most young transplanted trees) then it's plain, that but half the usual nourishment can be carried up, through the roots, on that account ; and a very much less proportion on account of the small hemisphere of earth, the new planted shortened roots occupy ; and on account of the loose position of the new turned earth, which touches the roots at first but in a few points. This (as well as experience) strongly evinces the great necessity of well watering new plantations."

This is a characteristic Halesian touch. These important quantitative velocity relations are immediately given practical application, and used to show the rational necessity of a practical rule, by showing that that rule expresses a principle derivative from the very general nature of plant life. A comparison between the water economy of Man and plant was now made possible by the possession of these figures, and is of the order, if due allowance is made for difference in bulk, of 17 to 1, so great is the flow of water through the life processes of plants.

The role of the leaves as the organs of perspiration has already been assumed, as can easily be seen, in the experiment quoted and the others of a similar kind with which Hales begins his study. Without this assumption those experiments could not have been conceived, and yet the assumption is, at that stage, quite untested. The next step then is to test the hypothesis of the essential role of the leaves. Experiment VII of Chapter 1 retails how " In *July* and *August* I cut off several branches of Apple-trees, Pear, Cherry, and Apricot-trees, two of a sort ; they were of several sizes from 3 to 6 feet long, with proportional lateral branches ; and the transverse cut of the largest part of their stems was about an inch diameter."

" I stripped the leaves off of one bough of each sort, and then set their stems in separate glasses, pouring in known quantities of water."

" The boughs with leaves on them imbibed some 15 ounces, some 20 ounces, 25 or 30 ounces in 12 hours day, more or less in proportion to the quantity of leaves they had ; and when I weighed them at night they were lighter than in the morning."

" While those without leaves imbibed but one ounce, and were heavier in the evening than in the morning, they having perspired little."

" The quantity imbibed by those with leaves decreased very much every day, the sap vessels being probably shrunk, at the transverse cut, and too much saturate with water, to let any more pass ; so that usually in 4 to 5 days the leaves faded and withered much."

" I repeated the same experiment with Elm-branches, Oak, Osier, Willow, Sallow, Aspen, Curran, Gooseberry ; and Philbert branches ; but none of these imbibed so much as the foregoing, and several sorts of the evergreens very much less."

Now we know how much water passes through the plant, and we know that it is somehow dependent on the presence or absence of *leaves*. But though necessity of leaves to water movement has been shown, the exact role of the leaves has yet to be elucidated. Merely to show that two phenomena are correlated is not to have discovered anything of much scientific interest. Science seeks for the answer to the questions : " Why ? ". Why are the leaves necessary to the water movement ? Hales' next set of experiments move tentatively towards an answer. Is the sap forced upwards by pressure and so exuded from the leaves ? Or is there some process of transpiration going on in the leaves which required that water be drawn up to them ? At least the beginnings to answers to these questions were provided by Hales in Experiments X and XI of Chapter 1. These accounts should be read in conjunction with a study of the accompanying illustrations (fig. 10 and fig. 11).

" *July* 27. I fixed an *Apple-branch m*, 3 feet long ½ inch diameter, full of leaves, and lateral shoots to the tube *t*, 7 feet long, ⅝ diameter. I filled the tube with water, and then immersed the whole branch as far as over the lowest end of the tube, into the vessel *u u* full of water."

" The water subsided 6 inches the first two hours (being the first filling of the sap vessels) and 6 inches the following night, 4 inches the next day ; and 2 + ½ the following night."

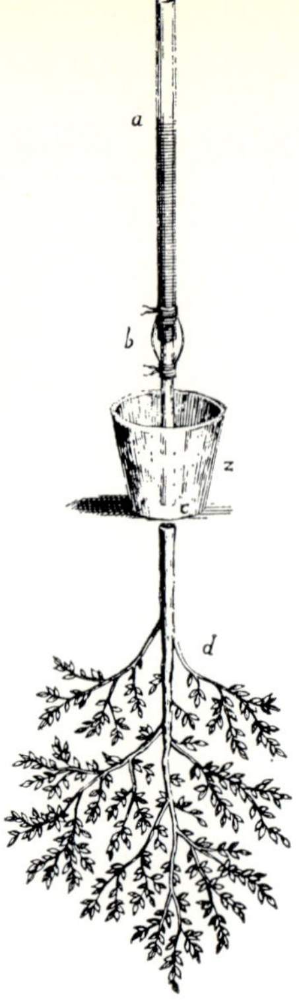

Fig. 10 Apparatus for measuring amount of water perspired under different conditions.

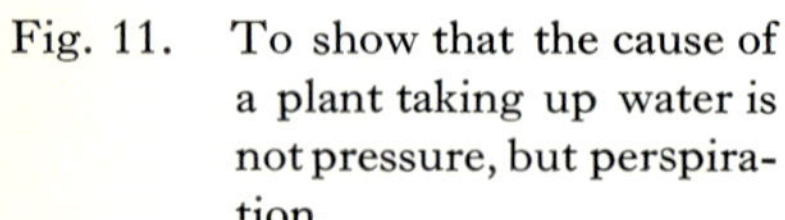

Fig. 11. To show that the cause of a plant taking up water is not pressure, but perspiration.

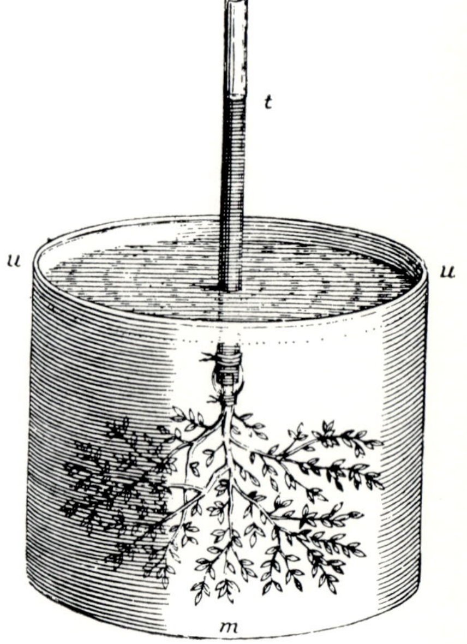

" The third day in the morning, I took the branch out of the water ; and hung it with the Tube affixed to it in the open air ; it imbibed this day $27+\frac{1}{2}$ inches in 12 hours."

" This Experiment shews the great power of perspiration ; since when the branch was immersed in the vessel of water, the 7 feet column of water in the tube, above the surface of the water, could drive very little through the leaves, till the branch was exposed to the open air."

" This also proves, that the perspiring matter of trees is rather actuated by warmth, and so exhaled, than protruded by the sap upwards."

" And this holds true in animals, for the perspiration in them is not always greatest in the greatest force of the blood ; but then often least of all, as in fevers."

" I have fixed many other branches in the same manner to long tubes, without immersing them in water ; which tubes, being filled with water, I could see precisely, by the descent of the water in the tube *t*, how fast it perspired off ; and how very little perspired in a rainy day, or when there were no leaves on the branches."

This series of experiments seems to show that it is the very process of perspiration from the leaves that is essential to the updrawing of water in the naturally growing plant, and certainly has nothing to do with the leaves being on the top of the plant. When the perspiration is prevented, water is not passed in quantity through the plant. In the course of this exposition Hales toys with the hypothesis that warmth has something to do with perspiration, comparing the rate of perspiration in animals and plants and conjecturing that the absolute pressure of blood or sap is not the effective agent. The proof of this conjecture is elegantly shown in his Experiment XI.

" *Aug*. 17. At 11 *a.m.* I cemented to the tube *ab* (fig. 10) 9 feet long, and $\frac{1}{2}$ inch diamter an *Apple-branch d* 5 feet long 6/8 inch diameter ; I poured water into the tube, which it imbibed plentifully, at the rate of 3 feet length of the tube in an hour. At 1 o'clock I cut off the branch at *c*, 13 inches below the glass-tube. To the bottom of the remaining stem I tied a glass cistern *z*, covered with ox-gut, to keep any water which dropped from the stem *c b* from evaporating. At the same time I set the branch *d r* which I had cut off in a known quantity of water in the vessel *x*, the branch in the vessel *x* imbibed 19 ounces of water, in 18 hours day and 12 hours night ; in which time only 6 ounces of water had passed through the stem *c b* which had a column of water 7 feet high pressing upon it all the time."

" This again shows the great power of perspiration ; to draw three times more water, in the same time, through the long slender parts of the branch *r* as was pressed through a larger stem *c b* of the same branch ; but 13 inches long, with 7 feet pressure of water upon it, in the tube *a b*."

We shall return to his study of the temperature conditions of the life of plants a little below, since this is concerned to explore a possible cause of " perspiration " rather than the phenomenon itself. The experiments

I have quoted so far show that the movement of water through the plant is not a simple pressure effect, and they also establish that the presence of leaves is a critical condition for normal transport of water through the plant. If only one could find *why* the leaves are necessary, one could know *how*, that is by what process the leaves draw up the water. But to complete the study of the passage of water through the system of the plant two further studies are needed. It is necessary to follow the water outside the plant, so as to establish that it is indeed water that is perspired. The complementary study will be to investigate the water situation in the earth, that is to follow the process backwards, as it were, into the earth from which the plant draws the water. The clarity of Hales' exposition of the experiments to collect the liquid transpired is such that we can do no better than read his own words. In Experiment XVII he says: "Having by many evident proofs in the foregoing experiments seen the great quantities of liquor that were imbibed and perspired by trees, I was desirous to try if I could get any of this perspiring matter; and in order to get it, I took several glass chemical retorts, *b a p* (fig. 12) and put the boughs of several sorts of

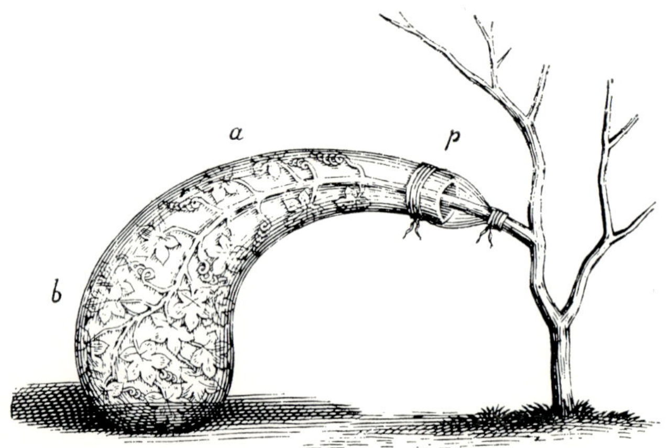

Fig. 12. Apparatus to collect perspired fluid.

trees, as they were growing with their leaves on, into the retorts, stopping up the mouth *p* of the retorts with bladder. By this means I got several ounces of the perspiring matter of Vines, Fig-trees, Apple-trees, Cherry-trees, Apricot and Peach-trees; Rue, Horseradish, Rhubarb, Parsnip, and Cabbage leaves: the liquor of them was very clear, nor could I discover any different taste in the several liquors: But if the retort stand exposed to the hot sun, the liquor will taste of the coddled leaves. Its specific gravity was nearly the same with that of common water; nor did I find many air bubbles in it, when placed in the exhausted receiver, which I expected to have found ; but when reserved in open viols, it stinks sooner than common water ; an argument that it is not pure water, but has some heterogeneous mixtures with it."

" I put also a large Sun-flower full blown and as it was growing, into the head of a glass-still, and put its rostrum into a bottle, by which there distilled a good quantity of liquor into the bottle. It will be very easy in the same manner to collect the perspirations of sweet scented Flowers, though the liquor will not long retain its grateful odour, but stink in a few days."

" This experiment would be very proper to begin the learned *Boerhaave's* clear and very rational chemical processes with, as being a degree more simple than his first process, the distillation in a cold still : For this is undisturbed nature's own method of distilling."

There is no doubt then that what is perspired is actually water, though the rough test of putrefaction shows that it cannot be absolutely pure. Water is drawn into the plant and water is perspired out. The next step is to investigate the source of this water, and to try to establish that it is possible that it all comes from the water stored in the earth. The average amount of water in a cubic foot of earth was found by Hales by digging out three separate and equal volumes of earth, one below the other. This sample would cover most of the volume of earth into which the roots of ordinary plants would normally penetrate. These samples were weighed immediately upon being dug up. They were let dry out until they were too dry to be capable of supporting plant life and then weighed again. The difference in weight was the amount of water lost in drying. Knowing the density of water it is then an easy matter to calculate the amount of water per unit volume of earth, and Hales found that each cubic foot of earth contained 7 lb of water which could be drawn off before the earth became too dry to support plant life, that is there was 0.112 kilogramme of water per litre of earth. Recollecting that Hales knew how much water a sunflower drew up from the earth in a day, it is easy to calculate, on the assumption that 112 litres of earth are penetrated by the roots and thus can serve as a source for water, that the plant can stay alive for 21 days and six hours without the water supply being replenished. Apart from deeper sources and springs in the ground there are two sources of water to replenish the volume of earth within which the plant grows. These are dew and rain. Now *they* must be investigated to see whether together they are sufficient to supply the water needs of the plant, or whether we need to postulate the replenishment of the supply by " moisture arising from below 15 inches (the depth of the roots) up into the earth occupied by the roots ".

To this end Hales measured the fall of dew, and then adding the amount of rain that generally falls, and *subtracting the amount lost in evaporation*, he was able to work out the actual available supply of water from above. It turns out that if we consider the balance between dew-fall and evaporation, that more evaporates from the surface of the earth. Indeed, more than 10 lb (4.5 kilogramme) of water will evaporate as a net loss, over the amount of dew fall, in the 21 days period that Hales uses to establish the water requirements of the sunflower. Now this shows that there must be other sources of water than the dew, since the earth never dries out to that extent. It involves a loss of water of 9 lb per cubic foot, (0.144 kilogramme per litre) in the 21 day period, and this is to be compared with Hales' experimental finding that in losing 7 lb per cubic foot (0.112 kilogramme per litre) the soil becomes too dry to support the life of plants. Two curious and unexpected pieces of information appear in the course of these experiments. It seems that the moister the earth, the more dew falls on it. It is also the case that the evaporation of water from the surface of earth in winter is much the same as in summer. The true source of water supply cannot be dew. Could it be rain ? The average rainfall is 22 inches a year (55 cm). Bringing all these facts together, we can easily calculate that the rain is an adequate source of water for the growth of plants. " Hence we find that 22 inches depth of rain in a year is sufficient for all the purposes of nature, in such flat countries as this about *Teddington* near *Hampton Court*. But in the hill countries, as in *Lancashire*, there falls 42 inches depth of rain-water ; from which deducting 7 inches for evaporation, there remains 35 inches depth of water for the springs. . . ."

In order to try to find out where the water actually went in a plant Hales devised a kind of experiment much used nowadays in biology, the *tracer* experiment. One knows that a plant or an animal absorbs some material but not exactly where it goes to in the plant. If a detectable " mark ", as it were, is made on the stuff, one could follow it through the body. Hales' idea was to use water which had something added, perfume, camphor and so on and then to see which parts of the plant smelled of this " trace element " after some time. Thus he hoped to follow the water through the plant. These days exactly similar experiments are done using radioactive substances which can be detected in those parts of the body to which they go by their radiation. Hales did find differential absorption of these substances. " I poured, " he says, " into a tube *l*, fixed to a golden Renate-tree, a quart of high rectified spirit of wine, camphorated, which quantity the stem imbibed in 3 hours space ; this killed one half of the tree : this I did to try if I could give a flavour of camphor to the apples which were in great plenty on the branch. I could not perceive any alteration in the taste of the apples, though they hung several weeks after ; but the smell of the camphor was very strong in the stalks of the leaves, and in every part of the dead branch."

" I made the same experiment of a vine, with strongly-scented orange-flower-water ; the event was the same, it did not penetrate into the grapes, but very sensibly into the wood and stalks of the leaves."

" I repeated the same experiment on two distant branches of a large Catherine pear-tree, with strong decoctions of sassafras, and of flowers, about 30 days before the pears were ripe ; but I could not perceive any taste of the decoctions in the pears."

" Though in all these cases the sap-vessels of the stem were strongly impregnated with a good quantity of these liquors ; yet the capillary vessels near the fruit were so fine, that they changed the texture of, and assimilated to their own substance those high tasted and perfumed liquors ; in the same manner as grafts and buds change the very different sap of the stock to that of their own specific nature."

" This experiment may safely be repeated with well scented and perfumed common water, which trees will imbibe at *l l* without any danger of killing them."

Quite apart of this being the first trace-element experiment in history it has other points of interest. Hales' discussion of the reasons why the fruit were not flavoured by the trace-substance refers to views about the nature of material stuff that were current in his day. The word " texture " does not mean the outward feel or appearance of something, as we use it today. Rather it means the arrangement of the particles or corpuscles which go to make up a thing. Differences in texture between things were differences in the arrangement of their minute parts, some arranged in cubes, some in tetrahedra and so on. The different textures of bodies were responsible for their different appearances, such as being rough or smooth, shiny or matt, and were also supposed to be responsible for differences in colour between things. Locke, the philosopher, for example, explained that the differences between black things and white things was due to " the difference arrangements of the particles in their superficies ". Differences in taste and smell were also explained as the effects of different arrangments of corpuscles in things, or differences in texture. It was a cornerstone of Robert Boyle's ideas on nature that all the appearances of things to us were the effects of the " bulk, figure (i.e. shape), texture and motion of the insensible parts ". Locke even copied the very phrase from Boyle, and most other scientists and philosophers of the time, such as Newton, agreed with this general idea. Differences in the motion or speed of vibration of the parts of things were responsible for our feelings of different degrees of warmth and so one. It is evident that Hales shared the " corpuscularian philosophy " of his time. Boyle, in a series of famous experiments, set out to show that all the qualities and forms that we saw in things were really the effects on us of differences in the arrangements and motions of the insensible parts, that is of parts too small to be seen or touched separately. He pointed out that simply by dividing up a thing into smaller parts one could change its colour, as for instance when we see that ground glass is white while a whole piece of glass is transparent. He also pointed out how the quality of white of egg changes as it is whisked, passing from a clear watery liquid to a sticky white substance. He also noticed how the various qualities of a chick are produced by the " cicatricula ", or egg spot, out of nothing but yoke and white of egg, neither of which, for example, has the springiness of the chick's tendons. He argued that in all the processes of nature, and in all the efforts of chemists in the laboratory, there were only breakings-up and rearrangements going on, that is, really there were only mechanical changes, changes in " bulk, figure, texture and motion " taking place. So every change in the qualities of things as seen by us, must be due to such mechanical changes, and so since mechanical changes can only change things mechanical, the cause of the appearance of the qualities themselves must be the same.

How does the moisture get from the deeper parts of the soil to plants ? After all, it is not always raining, and it is easy to observe that the plant does not progressively dry out the soil around it. Something causes the moisture to rise and to be continually replenished. Hales sought an explanation for this in the effect of heat on the water in the earth. To investigate this phenomenon he arranged several thermometers in the earth, each longer than the other, so that they could be embedded in the earth to different depths and all be read at the surface. He studied the temperature of the earth at different seasons of the year. He found that there was a falling-off in temperature as the depth increased, and that down to a certain depth there was a marked effect of the Sun's warmth upon the earth, but that at lower points, though the temperature rose and fell with the seasons, the temperature stayed fairly steady during the day. These facts, coupled with the corpuscularian theory of matter, gave him his theory.

" Now, so considerable a heat of the Sun, at two feet depth, under the earth's surface, must needs have a strong influence in raising the moisture at that and greater depths ; whereby a very great and continual wreak must always be ascending, during the warm summer season, by night as well as day ; for the heat at two feet depth is nearly the same night and day : The impulse of the Sun-beams giving the moisture of the earth a brisk and undulating motion, which watery particles, when separated and rarefied by heat, do ascend in the form of vapour : And the vigour of warm and confined vapour, (such as is that which is 1, 2 or 3 feet deep in the earth) must be very considerable, so as to penetrate the roots with some vigour ; as we may reasonably suppose, from the vast force of vapour in *Æolipiles*, in the digester of bones, and the engine to raise water by fire." In fact, according to Hales, the heat of the Sun causes the water in the earth to evaporate, and then being confined in the earth, this vapour increases in pressure, forcing its way upward, and passing more readily into the roots. " 'Tis therefore probable," he says, " that the roots of trees and plants are thus, by means of the Sun's warmth, constantly irrigated with fresh supplies of moisture ; which by the same means, insinuates itself with some vigour into the roots. For if the moisture of the earth were not thus actuated, the roots must then receive all their nourishment merely by imbibing the next adjoining moisture from the earth ; and consequently the shell of the earth, next to the surface of the roots, would always be considerably drier the nearer it is to the root; which I have not observed to be so." The complementary retardation of growth by cold is amply instanced by Hales, in comparing the years 1723–25.

Having established the general water economy of the plant in this first set of experiments Hales then set out to investigate the pressure of sap within the plant. Though plants have no "engine, which, by its

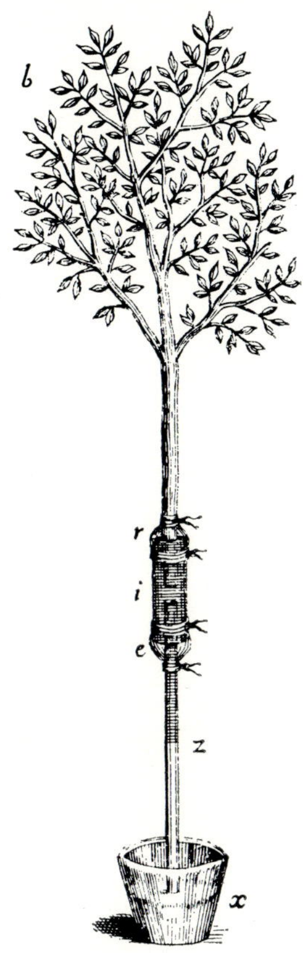

Fig. 13. The Aqueo-mercurial gauge.

alternate dilatations and contractions", pushes round the blood as in animals, nevertheless there is a movement of sap within the plant. But in order to investigate this a suitable instrument was required. Hales devised what he calls an "Aqueo-mercurial gauge". The construction of this gauge and the necessary corrections to be made in using its results is described in Experiment XXII. "*May* 25, I cut off a branch of a young thriving *Apple-tree b* (fig. 13) about 3 feet long, with lateral branches; the diameter of the transverse cut *i*, where it was cut off, was

$\frac{3}{4}$ of an inch : The great end of this branch I put into the cylindrical glass *e r*, which was an inch diameter within, and eight inches long."

" I then cemented fast the joint *r*, first folding a strap of sheeps skin round the stem, so as to make it fit well to the tube at *r* ; then I cemented fast the joint with a mixture of bees-wax and turpentine melted together in such a proportion, as to make a very stiff clammy paste when cold, and over the cement I folded several times wet bladders, binding it firm with Packthread."

" At the lower end of the large tube *e* was cemented, on a lesser tube *z e*, $\frac{1}{4}$ inch diameter in bore, and 18 inches long : The substance of this tube ought to be a full $\frac{3}{4}$ of an inch thick, else it will too easily break in making this experiment."

" These two tubes were cemented together at *e*, first with common hard brick-dust cement to keep the tubes firm to each other ; but this hard cement would, by the different dilatations and contractions of the glass and cement, separate from the glass in hot weather, so as to let in air ; to prevent which inconvenience. I further secured the joint with the cement of Bees-wax and Turpentine, binding a wet bladder over all."

" When the branch was thus fixed, I turned it downwards, and the glass tube upwards, and then filled both tubes full of water ; upon which I immediately applied the end of my finger to close up the end of the small tube, and immersed it as fast as I could in the glass cistern *x*, which was full of mercury and water."

" When the branch was now uppermost, and placed as in this figure, then the lower end of the branch was immersed 6 inches in water, *viz*, from *r* to *i*."

" Which water was imbibed by the branch, at its transverse cut *i* ; and as water ascended up the sap vessels of the branch, so the mercury ascended up the tube *e z* from the cistern *x* ; so as in half an hour's time the mercury was risen 5 inches and $\frac{3}{4}$ inches high up to *z*."

In this way Hales constructed and employed his gauge, which served both as a reservoir of water for the plant to imbibe and as a measuring device for pressure. You will notice the care with which the gauge was constructed, and Hales' clear understanding of the various weaknesses that are possible in the construction. Sound construction is one requirement for instruments, but there is another of equal importance, namely the correction for errors which are an inherent part of the experimental method. In the use of this gauge Hales realized that there was an inherent source of error, namely the presence of air, which appears in the tube *r e* from the end of the wood. As the wood imbibes water it gives out air. And this accumulates in the tube, affecting the height of the mercury and so interfering with the measurement of pressure by means of the reading of the height of the mercury drawn up. It is part of the mastery of Hales as an experimentalist that he is constantly on the lookout for such sources of error. Once identified, the error can be overcome. Hales points out that " this height of the mercury did in some measure show the force with which the sap was imbibed, though not near the whole force ; for while the water was imbibing, the transverse cut of the branch was covered with innumerable little hemispheres of air, and many air bubbles issued out of the sap vessels, which air did in part fill the tube *e r*, as the water was drawn out of it ; so that the height of the mercury could only be proportionable to the excess of the quantity of water drawn off, above the quantity of air which issued out of the wood ".

" And if the quantity of air, which issued from the wood into the tube, had been equal to the quantity of water imbibed, then the mercury would not rise at all ; because there would be no room for it in the tube."

" But if 9 parts in 12 of the water be imbibed by the branch, and in the mean time but 3 such parts of air issue into the tube, then the mercury must needs rise near 6 inches, and so proportionably in different cases."

We now have the instrument and know how to correct it. And with it Hales proceeded to determine the force or power of imbibing of a great many different plants and trees. He found considerable differences in the power of plants to imbibe water, noting in particular that it seems to be very much less amongst evergreens. He gives no explanation of the fact, though it is in fact explicable by one of his own most important discoveries, the importance of evaporation (perspiration) from the leaves as a factor in the drawing up of water. In general the rate of perspiration from evergreens is less than from broad-leafed plants, so the power of imbibing will also be less.

The water/mercury gauge showed that there is a power of imbibing water in both roots and in the stems of branches, as one would expect. One might also expect that since the presence or absence of leaves on a branch affect the rate and total quantity of water imbibed by a branch, they will also be a critical factor in the pressure or suction by which the plant draws up water. In Experiment XXVIII Hales measured the power of imbibing of an apple branch with no leaves and found that it very quickly vanished. In a more powerful experiment, (Experiment XXX) he compared the suctions of two very nearly identical apple branches, one with leaves the other without. Marked differences were observed in the height of the mercury, in one case a difference of 12 times. The absence of leaves then affects rate, quantity and suction of water, and Hales saw this was because, in the absence of leaves, the perspiration from the plant is drastically reduced.

Now this rate of perspiration, which is of such importance, is also affected, Hales discovered, by the presence or absence of sunlight, and consequently we should expect the power of suction to be similarly dependent upon the amount of sunlight falling on the plant. Indeed, in most of his experiments, Hales observed " that the mercury rose highest, when the sun was very clear and warm ; and towards evening it would subside 3 or 4 inches, and rise again the next day as it grew warm ". He notices also that as the days pass the greatest suction steadily declines through the deterioration of the vessels within the plant.

In Experiment XXIV Hales took account of a phenomenon only recently rediscovered. When a branch is cut, the vessels which are full of water drain to a certain extent. These vessels are crossed by transverse membranes with minute pores, which allow the passage of water, but close up when the water drains out, and so prevent the passage of air. In these circumstances an " air lock " can form in the stem and very much affect the passage of water, and so the effective suction. To overcome this Hales sucked out the air before quickly plugging the cut end into water, whence the suction was as normally expected. " But though " he says, " the quantity of those air bubbles thus sucked out, was but small ; yet in this and many other experiments, I found that after much suction, the water was imbibed by the branch, much more greedily, and in much greater quantity than the bulk of the air was, which was sucked out. Probably therefore, these air bubbles, when in the sap vessels, do stop the free ascent of the water, as in the case of little portions of air got between the water in capillary glass tubes."

Is there a preferred *direction* in which the water must pass ? This can be tested by seeing if it makes any difference whether the " great " end or the " small " end of the branch be immersed in the water of the gauge. In fact Hales found that either end of a branch will asborb water, provided the branch has lateral twigs with leaves. A powerful suction is produced, raising the level of the mercury. " Branches will strongly imbibe from the small end immersed in water to the great end ; as well as from the great end immersed in water to the small end." There is no preferred direction.

There can be no doubt of the role of the leaves in the process of water transport within the plant. But might there not be some effect too to be derived from the other external feature of the branches, namely the bark ? In a characteristically straightforward experiment Hales stripped the bark off two branches, fixed them in such a way that one was imbibing water from the small end, and the other from the great end. Both affected the gauge equally, and in the normal sort of way. So whether the branches have bark or not, is not a factor of any importance in their power to imbibe water. Thus the fact of the suction and the conditions under which it is exercised, and its relative magnitude in various plants and under various conditions, has been established. But Hales is a scientist, and he wants to know *why* the facts are as they are. He wants to know the mechanism of the suction and transport of water. After all there is no " engine " for pumping it around. The science of Hales' day, as we have seen already, was built upon the idea that material things were ultimately composed of corpuscles, small, indeed minute parts. The explanation of what was seen in nature was to be sought in the way the arrangements of these parts changed. In addition to this basic idea it was also believed that there were various fundamental forces at work which acted between corpuscles. Usually these were reduced to the relation of contact between corpuscles which actually struck one another, and attraction between those which were some distance apart. Gravity was the most striking and obvious example of an attraction between bodies which were not in contact. We have already explored the origin of this idea of attraction in the magical tradition which was so instrumental in producing new concepts in the very early period of the rise of modern science. Hales, too, found attraction attractive. " We see," he says, " in the Experiments of this chapter, many instances of the great efficacy of attraction ; that Universal principle which is so operative in all the very different works of nature ; and is most eminently so in vegetables, all whose minutest parts are curiously arranged in such order, as is best adapted by their united force, to attract proper nourishment." Various examples of the power of substances to attract water catch his attention, for example, wood ashes packed into a tube and red lead too will imbibe water and cause the aqueo-mercurial gauge to register a suction. To justify his generalization of the principle he quotes Query 31 from the *Opticks* in which Newton observes that " the water rises up to this height, by the action only of those particles of the ashes which are upon the surface of the elevated water ; the particles which are within the water, attracting or repelling it as much downwards as upwards ; and therefore the action of the particles is very strong : But the particles of the ashes being not so dense and close together as those of glass, their action is not so strong as that of glass, which keeps quick-silver suspended to the height of 60 or 70 inches, and therefore acts with a force which would keep water suspended above 60 feet ". Newton goes on to apply this idea more generally. " By the same principle," he says, " a sponge sucks in water, and the glands in the bodies of animals, according to their several natures and dispositions, suck in various juices from the blood." Hales extends the application of the principle still further. " And by the same principle it is, that we see in the preceding Experiments plants imbibe moisture so vigorously up their fine capillary vessels ; which moisture, as it is carried off in perspiration, (by the action of warmth,) thereby gives the sap vessels liberty to be almost continually attracting fresh supplies, which they could not do, if they were fully saturate with moisture : For without perspiration the sap must necessarily stagnate, not withstanding the sap vessels are so curiously adapted by their exceeding fineness, to raise the sap to great heights, in reciprocal proportion to their very minute diameters."

The third main problem about the sap was the question of its circulation. In part this problem was posed by the evident analogy between sap and blood, and in part by traditional ideas about the upward movement of water, and in part by reflection on the consequences of the ideas of Malpighi and Grew as to the role of the leaves in the production and elaboration of food material in the plant, which must then be passed by some means to parts other than the leaves. The commonest theory held in Hales' time supposed that there was some sort of circulation, though whether there was pulsation or not was an open question, and that the sap moved *up* in the *inner*most part of the stem, and *down* in the outer layers.

Hales first showed that water can pass in any direction around the plant, provided there are some leaves, the perspiration from which can provide the motive force, so to say. For instance, he immersed some of the leaves of a branch in water, and showed that the plant could imbibe water, and that in considerable quantities through the leaves. In

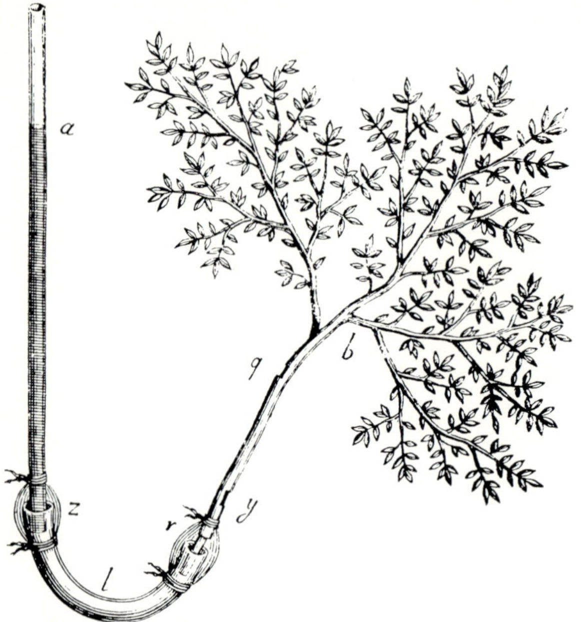

Fig. 14. Experiment on the circulation of the sap.

nature, he believed, plants were capable of directly absorbing dew and rain through the leaves, which accounted for their increase in weight on dewy nights. He also pointed out the advice of the famous gardener Mr. Miller, to wash the leaves and bark of newly planted trees. Evidently then water can pass either way through the leaves and branches of a plant.

In two further experiments Hales disposed for all time of the theory that sap always moved up in the inner part of the stem and down in the outer. In Experiment XLIII he demolishes the theory that the sap descends by the outermost part of the wood or the bark. On 20 August he says, " at 1 *p.m.* I took an *Apple-branch b* (fig. 14) nine feet long, $1 + \frac{3}{4}$ inch diameter, with proportional lateral branches, I cemented it fast to the tube *a*, by means of the leaden syphon *l*: but first I cut away the bark, and last years ringlet of wood, for 3 inches length to *r*. I then filled the tube with water, which was 22 feet long, and $\frac{1}{2}$ inch diameter, having first cut a gap at *y* through the bark, and last year's wood 12 inches from the lower end of the stem: the water was very freely imbibed, *viz.* at the rate of $3 + \frac{1}{2}$ inches in a minute. In half an hour's time I could plainly perceive the lower part of the gap *y* to be moister than before; when at the same time upper part of the wound looked white and dry ".

" Now in this case the water must necessarily ascend from the tube, through the innermost wood, because the last year's wood was cut away, for 3 inches length all round the stem; and consequently, if the sap in its natural course descended by the last year's ringlet of wood, and between that and the bark (as many have thought) the water should have descended by the last year's wood, or the bark, and so have first moistened the upper part of the gap *y*; but on the contrary, the lower part was moistened, and not the upper part ". So it is *possible* for a branch to take water up through its inner part, but when it does there is no corresponding downward movement in the outer part of the wood, or the bark. It follows therefore that there is no circulation. In a supplementary experiment (XLV) Hales stripped the bark off only one branch of a set of branches of similar size, all of which had their great ends immersed in water. If the sap descended by the bark or outer portion of wood then, he supposed, the leaves on the branch which had had some bark removed would stay green longer because they would not lose sap so readily as the others. But the whole lot lost their leaves at about the same time.

In Experiment XLIII Hales had observed that it was the *lower* end of a notch in the bark that bled sap. In another series of experiments he confirmed this phenomenon in other cases, from which he concluded that " 'tis probable that the sap ascends between the bark and wood, as well as by other parts ". In short, at least during the day, the general motion of sap is upwards through the whole cross section of the stem and branches of the plant, which plays an essentially passive role in the total movement of water, which is drawn off by reason of the tension caused by perspiration. This view is held to be essentially correct by the best modern authorities, though it is not thought that capillarity is wholly responsible for the transfer of the tension from the leaves downwards. As Hales says: " When the sap has first passed through that thick and fine strainer, the bark of the root, we then find it in greatest quantities, in the most lax part, between the bark and wood, and *that* the same through the whole tree. And if in the early spring, the Oak and several other trees were to be examined near the top and bottom, when the sap first begins to move, so as to make the bark easily run, or peel off, I believe it would be found that the lower bark is first moistened whereas the bark at the top branches ought first to be moistened, if the sap descends by the bark ; As to the Vine, I am pretty well assured that the lower bark is first moistened ". Not only is it untrue that there is a circulation but it is at any rate implausible. If the sap had to go up and down and around the tree before it was transpired from the leaves it would have to move with a very great " celerity ", considering how much a plant draws up and perspires in a day. The less efficient kind of movement in the plant compared with the circulation of an animal is made up for, Hales believes, by the great amount of water that the plant uses. He refers again to his own earlier discovery that in the case of the sunflower the plant uses, bulk for bulk, 17 times as much water in a day as a man. Plant life is essentially passive. As Hales puts it : " Nature's great aim in vegetables being only that the vegetables life be carried on and maintained, there was no occasion to give its sap the rapid motion, which was necessary for the blood of animals ".

In summary, then, the results of this masterly series of experiments into the life of plants are that " in vegetables we can discover no other cause of the sap's motion, but the strong attraction of the capillary sap vessels, assisted by the brisk undulations and vibrations, caused by the Sun's warmth, whereby the sap is carried up to the top of the tallest trees, and is there perspired off through the leaves : But when the surface of the tree is greatly diminished by the loss of its leaves, then also the perspiration and motion of the sap is proportionably diminished, as is plain from many of the foregoing Experiments : So that the ascending velocity of the sap is principally accelerated by the plentiful perspiration of the leaves, thereby making room for the fine capillary vessels to exert their vastly attracting power, which perspiration is effected by the brisk rarifying vibrations of warmth : A power that does not seem to be any ways well adapted, to make the sap descend from the tops of vegetables by different vessels to the root ". As to the hypothesis of circulation : " I think we have," says Hales, " sufficient ground to believe that there is no circulation of the sap in vegetables ; notwithstanding many ingenious persons have been induced to think there was, from several curious observations and experiments which evidently prove, that the sap does in some measure recede from the top towards the lower parts of plants, when they were with good probability of reason induced to think that the sap circulated."

" The likeliest method effectually and convincingly to determine this difficulty, whether the sap circulates or not, would be by ocular inspection, if that could be attained : And I see no reason we have to despair of it, since by the great quantities imbibed and perspired, we have good ground to think, that the progressive motion of the sap is considerable in the largest sap vessels of the transparent stems of leaves : And if our eyes, assisted with microscopes, could come at this desirable sight, I make no doubt but that we should see the sap, which was progressive in the heat of the day, would on the coming on of the cool evening, and the falling dew, be retrograde in the same vessels."

Further reading

It is hoped that those who use this book will have a copy of Stephen Hales' *Vegetable Staticks*, obtainable in a good paperback edition in the Oldbourne Science Library, edited by Dr M. A. Hoskin. This chapter should be read in conjunction too with a good modern textbook of plant science, such as *The Living Plant*, by Alan Brook, Edinburgh University Press.

Practical work

Most of Stephen Hales' experiments can be done with very simple equipment, so the selection of experiments suggested in this section should not be taken to exclude a different choice from the many described in *Vegetable Staticks*. The use of rubber tubing instead of cement, wet bladders, sheep skin and pack thread will greatly simplify the setting up of the apparatus.

1. The water economy of a simple cabbage plant can be worked out relatively easily. The class should be equipped with some cabbage plants growing in pots, which are well grown but have not yet started to " heart ". A sheet of polythene should be fixed over the top of the pot and wrapped round the stem. The plant and pot should be weighed every day, and after each weighing a weighed quantity of water should be added to the soil in the pot. From these results the average amount of water perspired in a day can be calculated. The next step is to form an estimate of the leaf area, the root area and the cross-sectional area of the stem.

To measure the surface of the leaves Hales made a grid of squares and placed it over the leaf, and counted the number of squares to cover the leaf. This can be done on a sheet of transparent plastic, ruling it up into 1 cm squares. From this result the rate of perspiration in cubic centimetres per square centimetre can be calculated. Next the total length of the roots can be measured and their average diameter estimated. It is then an easy matter to calculate their total area, assuming them to be uniform cylinders. The rate of absorption can then be calculated. Finally, from a measurement of the diameter of the stem, its cross section can be worked out and the rate of flow of water in it calculated.

2. The role of the leaves in the water economy of the plant can be shown by comparison of the rates of water absorption of two plants or branches of roughly similar size and the same species, one with the leaves on and the other without leaves. A number of comparative experiments can be done, simply immersing the cut ends of branches in water, and measuring the amount of water lost in a day. To avoid complications with water loss through evaporation polythene sheet can be used to seal off the top of the vessel in which the branches are put.

3. In the next series of experiments we shall test the " perspiration " theory. For this experiment we shall need a leafy branch, connected by rubber tubing to a glass tube of roughly the same diameter as the

branch, and a metre or so long. Hales used a tube over 2 metres long for this experiment but this length of tube is rather awkward to handle. The branch should be hung " head down ", as it were, with the tube vertical above it.

i. In the first experiment the tube is filled with water and the branch hung upside down in the open air. The amount of water taken through the branch can be measured by measuring the fall in the level of water in the tube in, say, three hours. The branch is now immersed in a tub of water so that all the leaves are under water. The difference in the rate of water absorption can be measured by comparing the amount the level of water falls in the tube in a comparable time. The leaves cannot perspire under water. What can you conclude from this experiment about the need for free perspiration from the leaves ?

ii. Different atmospheric conditions may affect the rate of perspiration. Try to find out the effect of wind, of sun and of a very humid atmosphere on the rate of perspiration by hanging out a branch under differing conditions and measuring the amount the level in the tube falls in a given time.

We can conclude from experiment (*i*) that the mere pressure of water in the tube is not enough to cause sap to pass out through the leaves. But this can be shown even more strikingly by another experiment.

iii. If we take another branch prepared in the same way as for the other experiments we can measure the amount of water it takes up in a given time with its leaves intact. If we now saw off the branch, leaving about 20 cm only attached to the tube, if it were pressure only which was responsible for the uptake of water by the plant, the water in the tube should force itself through the piece of branch. Find out whether it does by measuring the fall in water level in the given time after the branch has be ensawn.

4. The next step is to find out what it is that is perspired by plants. Take a large glass jar and put the whole of a leafy branch into it, sealing it in with polythene sheet. After some time some liquid will condense inside the jar and collect at the bottom. The same can be done with a cabbage plant, taking care not to break the stem. Provided the experiment is not done in the hot Sun the liquor so condensed can be tested by taste. What is it ? By weighing 1 cm^3 of it the specific gravity can be found. Does this confirm your taste hypothesis ? Leave some of the liquor in an open jar for several days. What can you conclude from its smell about its purity ?

5. Take equal volumes of earth from the surface of the ground, from 25 cm depth and another from 50 cm depth. Weigh them immediately. Then allow them to dry out for a week or so. Re-weigh. How much water did they contain per litre ? Convert Hales' figures to the same system of units and compare your average with his.

6. Hales believed that the warmth of the earth was responsible for the diffusion of water up into the places where plants drew their supply.

To find out how the temperature of the earth varies with depth, and varies with the weather, take four metal tubes, large enough to allow a thermometer tied to a string to be dropped down them. One should be 10 cm long, the next 25 cm, then 50 cm and the last 1 metre. With the end stopped up to prevent the tube being choked with earth they should be driven into the ground completely. Thermometers secured with string can be dropped down, and the temperature of the earth at these depths determined regularly, so that variations with weather and with the season can be found. Variation with depth and variation at a given depth with weather and season can then be plotted on graphs. Compare your results with a similar plot of Hales' temperatures. You will have to use your ingenuity to convert Hales' degrees to centigrade.

7. A water/mercury suction gauge can be made from chemical glassware and rubber tubing. It is essential that the wider part of the tube, *r i* in Hales' drawing, should be sufficiently large to contain an adequate reservoir of water for the tree to imbibe. It should be at least 5 cm in diameter and not less than 15 cm long. With the gauge measure the suction of a root of a tree or fairly substantial shrub. In a similar way the suction of a branch can be measured. It might be of interest to do an experiment that Hales himself did not perform, that is to find the suction of a whole cabbage plant, first by immersing the roots in the gauge, making sure that a good seal has been made around the stem of the plant and with the gauge. Then the roots can be cut off and the stem inserted in the gauge and the consequent suction measured.

8. Comparative suctions should now be measured between broad-leafed plants and evergreens, for example, between roughly similarly sized branches of oak and of pine. Further comparisons should be made between branches with and without their leaves.

9. Finally we must study the problem of circulation of the sap. We must try to discover whether the sap passes up or down the outer layers of the stem. For this experiment a branch with plenty of leaves should be prepared with a square notch cut in the bark and *just* into the wood, about 5 cm in length, about 15 cm above the cut end of the branch. The end of the branch should then be put in water. Which end of the notch first becomes moist ? What do you conclude about the direction of movement of the sap in the outer layers of the stem ?

CHAPTER 9

the air

THE importance of air to the growth of plants had already been grasped by Malpighi and Grew. The identification of a constituent of the air which supported life had been made by Mayow. Clearly air is vital to plants, and some of it is absorbed by them, or *fixed* as early chemists put it. Hales set out to find out in detail the way the air was involved in the life of plants. Unfortunately he made, or rather one should say he subscribed to, an erroneous notion of the air which led to serious confusion and misinterpretation of experiments on his part. He was confused between what we might call the *constituents* of the air as separable parts, and the *qualities* of the air such as elasticity, which he believed was able to be affected and changed in various processes. When he found that after certain operations water rose up inside a vessel in a certain way, he was not clear whether this was because a certain quantity of elastic air had been removed, or because the elasticity of the air had been destroyed, or even whether the removal of the elastic constituent had caused the remaining air to lose its elasticity. He vacillates among all three of these views. Most of the time he keeps to the view that there is an elastic constituent, but at the point where it really matters in the interpretation of the experiments on combustion and respiration, he supposes that these processes destroy the elasticity of the air. Had he confirmed and followed up Mayow's discovery of oxygen, that is a life and combustion supporting *constituent* of air, which is extracted by combustion and breathing, he would have anticipated Lavoisier in every particular.

It is certainly clear that Hales was well aware of the necessity of air to life. " It is well known," he says, " that air is a fine elastic fluid, with particles of very different natures floating in it, whereby it is admirably fitted by the great author of nature, to be the breath of life, of vegetables, as well as of animals, without which they can no more live, nor thrive than animals can."

Hales believed that air was taken into the plant through pores found in the roots, the bark and the leaves. He cites Grew's observations through the microscope of pores in the bark and stomata in the leaves. To confirm the hypothesis that air is indeed capable of being taken in at these places, he set up an experiment (fig. 15).

" I took a cylinder of Birch," he says, " with the bark on, 16 inches and $\frac{3}{4}$ diameter, and cemented it fast at *z* ; to the hole in the top of the air pump receiver *p p*, setting the lower end of it in the cistern of water *x*; the upper end of it at *n* was well closed up with melted cement."

" I then drew the air out of the receiver, upon which innumerable air bubbles issued continually out of the stick into the water *x*. I kept the receiver exhausted all that day, and the following night, and till the next day at noon, the air all the while issuing into the water *x*: I continued it thus long in this state, that I might be well assured, that the air must

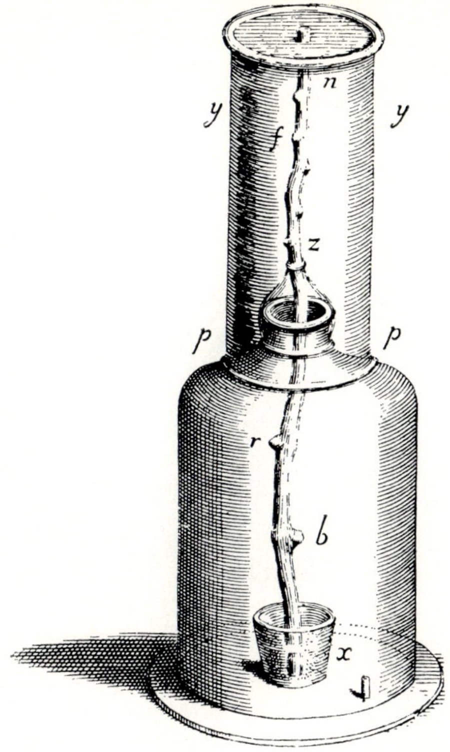

Fig. 15. To show that a plant can absorb air.

pass in through the bark, to supply that great and long flux of air at *x*. I then cemented up 5 old eyes in the stick, between *z* and *n*, where little shoots had formerly been, but were now perished, yet the air still continued to flow freely at *x*." To make sure that there was no other unnoticed source of air Hales fitted another vessel over the top part of the stick and filled it with water. The air soon ceased coming out of the lower part of the stick. Even with the water removed the air did not flow again, until the bark had been dried, thus suggesting very strongly that there were pores in the bark which had been stopped up by the water in some way.

Hales' conclusion has his usual uncanny perspicacity. " Whence it is very probable," he says, " that the air freely enters plants, not only with the principal fund of nourishment by the roots, but also through the surface of their trunks and leaves, especially at night, when they are changed from a perspiring to a strongly imbibing state."

It is now established that air can enter the living plant. It is necessary to establish that it is fixed in plants, animals and minerals. If air is fixed in the substance of these things then it is an essential constituent of them. The proof that this is so can be had from two series of experiments. The first series tries to recover the fixed air by distillation of appropriate substances, driving off " air " by heating, and collecting

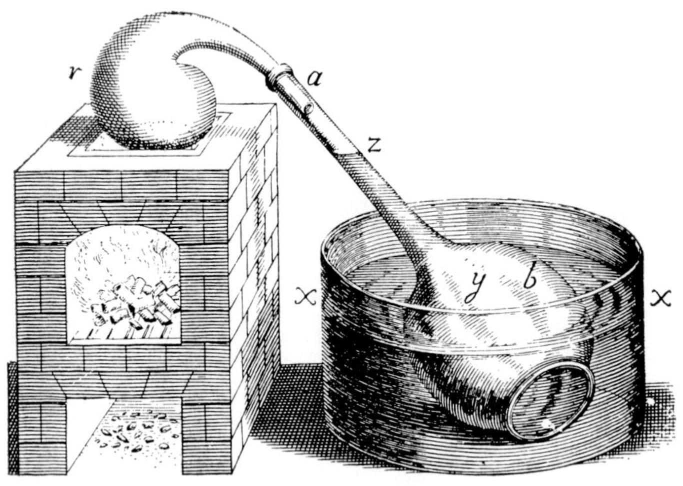

Fig. 16. Apparatus to release " fixed " air.

it and showing it to be elastic. The other way is by " fermentation " This is not fermentation as we understand it, rather it is a much broader idea. In Hales' time no clear distinction had been made between the phenomenon of the evolution of gas in such a process as brewing, and the evolution of gas in such a process as the action of a mineral or organic acid upon an alkali. All such processes were called fermentation. The fact that yeast is a living organism and that it produces the gas in the course of its life processes was only grasped 100 years later. So any case of the evolution of gas by putting together or mixing things is a fermentation.

The process of distillation is carefully described (see fig. 16). " In order to make an estimate of the quantity of Air, which arose from any body by distillation or fusion, I first put the matter which I intended to distill into the small retort *r* and then at *a* cemented fast to it the glass vessel *a b*, which was very capacious at *b*, with a hole in the bottom. I bound bladder over the cement which was tobacco-pipe clay and bean flour, well mixed with some hair, tieing over all four small sticks, which served as splinters to strengthen the joint ;" "... Matters being thus prepared, holding the retort uppermost, the open ended collection vessel was immersed in the large vessel *x x*, and the place *z*, to which the water rose, was marked with a wax thread. The retort was then brought up slowly to the fire and the collecting vessels screened from the heat." " The descent of the water in the [collecting vessel] showed the sums of the expansion of the Air, and of the matter which was distilling : The expansion of the Air alone, when the lower part of the retort was beginning to be red hot, was at a medium, nearly equal to the capacity of the retorts, so that it then took up a double space ; and in a white and almost melting heat, the Air took up a triple space or something more : for

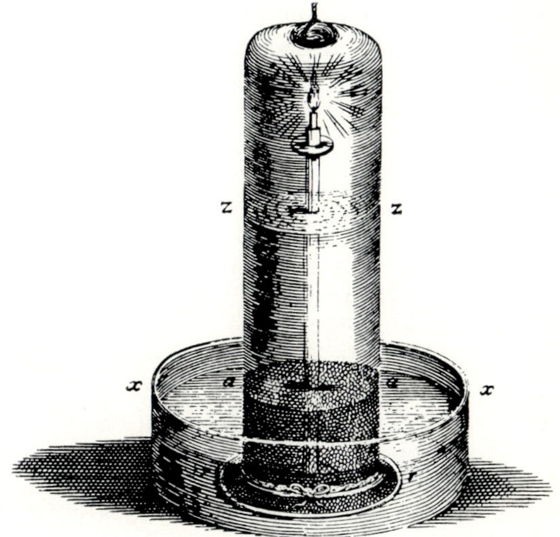

Fig. 17. To show the loss of elasticity caused by combustion.

which reason the least retorts are best for these Experiments. The expansion of the distilling bodies was sometimes very little, and sometimes many times greater than that of the Air in the retort, according to their different natures."

" When the matter was sufficiently distilled, the retort etc. was gradually removed from the fire, and when cool enough, was carried into another room, where there was no fire. When all was thoroughly cold, either the following day, or sometimes 3 or 4 days after, I marked the surface of the water *y*, where it then stood ; if the surface of the water was much below *z*, then the empty space between *y* and *z* showed how much Air was generated, or raised from a fixed to an elastic state, by the action of the fire in distillation : But if *y* the surface of the water was above *z*, the space between *z* and *y* ,which was filled with water, showed the quantity of Air which had been absorbed in the operation, i.e. was changed from a repelling elastic to a fixed state, by the strong attraction of other particles, which I therefore call absorbing."

For estimating the amount of air fixed or the amount of elasticity lost in combustion or respiration a different apparatus was required, " When I would make an estimate of the quantity of Air absorbed and fixed, or generated by the burning candle," says Hales, ". . . or by the breath of a living animal, etc. I first placed a high stand, or pedestal in the vessel full of water *x x* (see fig. 17) which pedestal reached a little higher than *z z*. On this pedestal I placed the candle, or living animal, and then whelmed over it the large inverted glass *z z a a*, which was suspended by a cord, so as to have its mouth *r r* three or four inches under water ; then with a syphon I sucked the Air out of the glass vessel till the water rose to *z z*. But when any noxious thing, as burning brimstone, aqua-fortis, or the like, were placed under the glass ; then by affixing to the syphon the nose of a large pair of bellows, whose wide sucking orifice was closed up, as the bellows were enlarged they drew the Air briskly out of the glass *z z a a* through the syphon ; the other leg of which syphon I immediately drew from under the glass vessel, marking the height of the water *z z*."

" When the materials on the pedestal generated Air, then the water would subside from *z z* to *a a* ,which space *z z a a* was equal to the quantity of Air generated : But when the materials destroyed part of the Air's elasticity, then water would rise from *a a* (the height that I in that case at first sucked it to) to *z z*, and the space *a a z z* was equal to the quantity of Air, whose elasticity was destroyed."

Hales invented a third sort of gas-collecting apparatus, which involved the novel principle of collection of the gas by the displacement of water, as in the modern water-bath method, of which this was the precursor. However, the point of the apparatus was the washing of the " Air " so as to remove " acid sulphurous fumes " which Hales supposed absorbed some of the air generated in the course of certain experiments, particularly the high temperature " distillation " of brimstone. The apparatus is illustrated in fig. 18. The need for and use of the apparatus is described in Experiments LXXVI and LXXVII.

The experiments to produce air by distillation, to unfix it as it were, comprised three series ; a series to produce air from animal products, a series using vegetable products, and a mineral series. A typical animal products experiment is XLIX. "A cubic inch of *Hog's blood*, distilled to dry scoria, produced thirty three cubic inches of Air, which Air did not arise till the white fumes ; which was plain to be seen by the great descent of the water at that time, in the receiver *a z y*." Amongst other animal products he distilled tallow, horn, and oyster shell. Typical of his vegetable product experiments was LVI in which he showed that " from 388 grain weight of *Indian Wheat*, which grew in my garden, but was not come to full maturity, was generated 270 cubic inches of air, the weight of which air was 77 grains, *viz.* $\frac{1}{4}$ of the weight of the wheat." Mustard seed, amber, tobacco and various oils were also distilled to produce permanent air." Similarly he distilled minerals like antimony

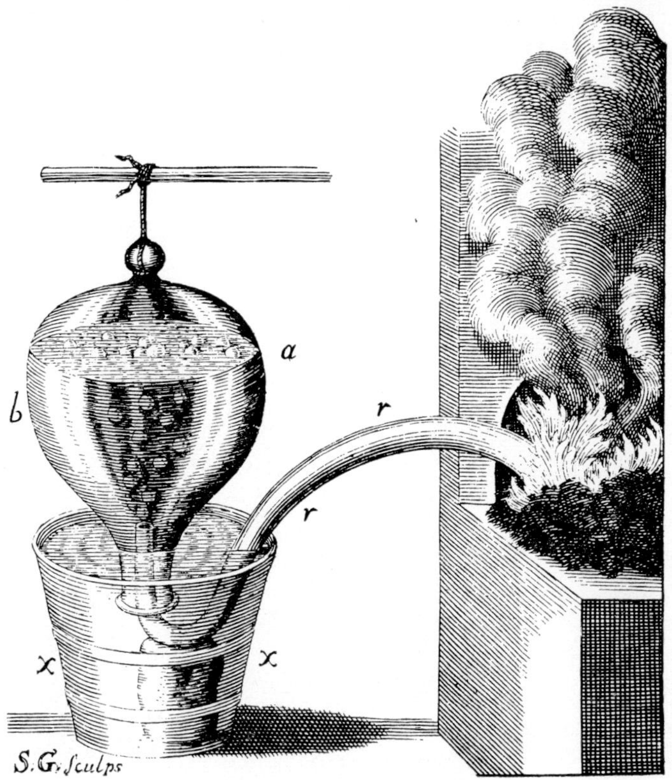

Fig. 18. Collecting " air " over water.

and sea-salt. He sums up his conclusions : " *As to animal substances*, a very considerable quantity of permanent air was produced by distillation, not only from the blood and fat, but also from the most solid parts of animals." " By the same means also, I found plenty of air might be obtained from *minerals*. And from vegetables too."

In the course of these experiments Hales discovered that when a substance is burnt it absorbs air and increases in weight. This was the discovery re-made by Lavoisier which transformed chemistry so many years later. Hales wholly failed to grasp the significance of his results. He just puts them down amongst other examples of the distillation of air. He says : " Two grains of *Phosphorus* easily melted at some distance from the fire, flamed and filled the retort with white fumes, it absorbed three cubic inches of air. A like quantity of *Phosphorus*, fired in a large receiver expanded into a space equal to sixty cubic inches, and absorbed 28 cubic inches of air : When 3 grains of *Phosphorus* were weighed, soon after it was burnt, it had lost half a grain of its weight ; but when two grains of *Phosphorus* were weighed, some hours after it was burnt, having run more *per deliquium* by absorbing the moisture of the air, it had increased a grain by weight ".

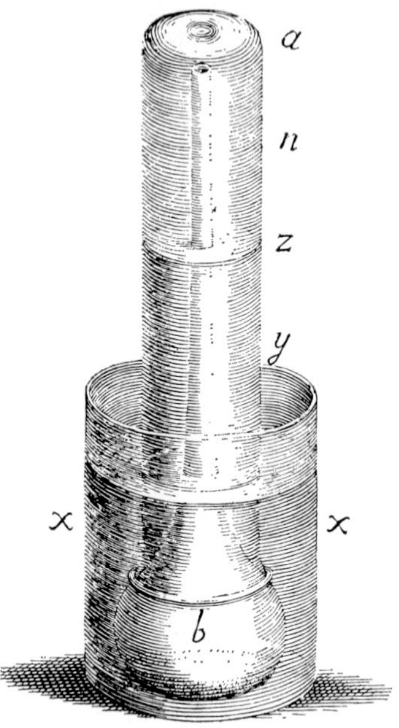

Fig. 19. Collection of gases by displacement of water.

In the next series of experiments Hales investigated the production of air by " fermentation ", that is both by chemical reaction and by organic action, that is by what we should nowadays call fermentation. The chemical reactions he studies are familiar ones. A typical example is Experiment LXXXIII. " In *February*," Hales says, " I poured on six cubic inches of powdered *Oystershell*, an equal quantity of common white-wine *Vinegar*. In 5 or 6 minutes it generated 17 cubic inches of air, and in some hours 12 cubic inches more, in all 29 cubic inches. In nine days it had slowly resorbed 21 cubic inches of air. The ninth day I poured warm water into the vessel *x x* (fig. 19) and the following day, when all was cool, I found that it had resorbed the remaining 8 cubic inches. Hence we see that warmth will sometimes promote a resorbing as well as a generating state, *viz.* by raising the resorbing fumes, as will appear more hereafter." Again Hales failed to grasp the significance of this experiment, and did not see the explanation for the resorbtion of some of the " air ".

Again, in studying genuine fermentation, he observed the resorbtion phenomenon. On " *March* the 2nd, 2 cubic inches of *Malaga Raisins*, with 18 cubic inches of *water* generated by the 16th of *April* 411 cubic inches of air; and then in 2 or 3 days it resorbed 35 cubic inches. From the 21st of *April* to the 16th of *May* it generated 78 cubic inches; after which to the 9th of *June* it continued in a resorbing state, so as to resorb 13 cubic inches: there was at this season many hot days, with much Thunder and Lightning, which destroys the air's elasticity; so there was generated in all 489 cubic inches of which 49 were resorbed. The liquor was at last very vapid ". In his conclusion on the results of genuine fermentation experiments we can see how far Hales was from any sort of clear conception of what was happening in the process. " We see from these Experiments," he says, " that in warm weather Wine and Ale do not turn vapid by imbibing air, but by fermenting and generating too much, whereby they are deprived of their enlivening principle, the air; for which reason these liquors are best preserved in cool cellars, whereby this active invigorating principle is kept within due bounds, which when they exceed, Wines are upon the fret and in danger of being spoiled ". By " vapid ", Hales means what we would call " flat ". It should be clear from this passage that he quite fails to distinguish beer losing what dissolved carbon dioxide it has and hence becoming flat, and wine starting to referment, and so generating new carbon dioxide in the course of turning vinegary. This " air " is fixed, and he does not have the conception of a liquid holding gas by virtue of its being dissolved in the liquid. He cannot distinguish that from fixation, which is chemical combination.

Having investigated the freeing of air which has been fixed already into animal, vegetable and mineral substances, Hales turned his attention to the process of fixation. He carried out a series of experiments designed to determine the effect on the air of burning, and the life processes of living things. He understood clearly that breathing and combustion had something in common, but he quite failed to keep clear the difference between absorbing and fixing a constituent of the air, and processes which destroyed its elasticity.

In Experiment CII he " fixed upon the pedestal under the inverted glass *z z a a*, (fig. 17) a piece of *Brown Paper*, which had been dipped in a solution of *Nitre*, and then well dried; I set fire to the paper by means of a burning glass: The *Nitre* detonized and burnt briskly for some time, till the glass *z z a a* was very full of thick fumes, which extinguished it. The expansion caused by the burning *Nitre* was equal to more than two quarts: When all was cool, there was near 80 cubic inches of new generated air, which arose from a small quantity of detonized *Nitre*; but the elasticity of this new air daily decreased, in the same manner as Mr. *Hauksbee* observed the air of fired Gun-powder to do, *Physicomechanical*

Experiments p. 83 so that he found 19 of 20 parts occupied by this air to be deserted in 18 days, and its space filled by the ascending water ; at which station it rested, continuing there for 8 days without alteration : And in like manner, I found that a considerable part of the air which was produced by fire in the distillation of several substances, did gradually lose its elasticity in a few days after the distillation was over ; but it was not so when I distilled air through water." Again we see the fatal confusion between loss of elasticity, a property of gases, and the loss of a component or constituent of a mixture of gases. For Hales it was really all the same air that was produced, so the rise in height of the water in the jar must be due to a decline in the power of the air to hold it out, that is a loss in elasticity. Instead, of course, the rise in the height of the water is due to the dissolving of the gas formed in combustion, either carbon dioxide or sulphur dioxide, in most of his experiments, in the water. And that of course is why the gas collected through water does not show this loss of " elasticity ", since in the passage through the water the carbon dioxide is already dissolved out. Again in his experiment on the burning candle (CVI) he remarks : " as the air cooled and condensed in the receiver, the water would continue rising above the mark, not only till all was cool, but for 20 or 30 hours after that, which height it kept, though it stood many days ; which shows that the air did not recover the elasticity which it had lost." This is in sharp contrast with Mayow's interpretation of the same and similar experiments many years earlier. Mayow was perfectly clear that the air was a mixture of various constituents, one of which, the *spiritus nitro-aereus*, was *removed* by combustion and breathing, to be held in combination or fixed in the ash or in the animal body which thus increased in weight. Mayow, however, himself failed to grasp that another gas, namely carbon dioxide, was produced in combustion and breathed out by animals. Yet he was much in advance of Hales in this matter who continued to believe that it was the elasticity of the air which was the essential vital feature of it and that air was a homogenous fluid. Hales was familiar with Mayow's work and on *May* 18 (probably of 1725 or 6) he repeated Mayow's experiment on breathing. He placed on a pedestal, " under the inverted glass *z z a a*, a full grown *Rat*. At first the water subsided a little, which was occasioned by the rarification of the air, caused by the heat of the Animal's body. But after a few minutes the water began to rise, and continued rising as long as the Rat lived, which was about 14 hours. The bulk of the air in which the Rat lived so many hours was 2024 cubic inches ; the quantity of elastic air which was absorbed was 73 cubic inches, above 1/27 part of the whole, nearly what was absorbed by a candle in the same vessel ". Notice that here he speaks of the removal of a *quantity* of elastic air.

With great boldness Hales performed one of the great classical experiments in human physiology, to find out the effect of human breathing on the air. "The following experiment will show," he says, " that the elasticity of the air is greatly destroyed by the *respiration of human lungs*, viz."

" I made a bladder very supple by wetting of it, and then cut off so much of the neck as would make a hole wide enough for the biggest end of a large fosset [tap] to enter, to which the bladder was bound fast. The bladder and fosset contained 74 cubic inches. Having blown up the bladder, I put the small end of the fosset into my mouth ; and at the same time pinched my nostrils close so that no air might pass that way, so that I could only breath to and fro the air contained in the bladder. In less than half a minute I found considerable difficulty in breathing, and was forced after that to fetch my breath very fast ; and at the end of the minute the suffocating uneasiness was so great that I was forced to take away the bladder from my mouth. Towards the end of the minute the bladder was become so flaccid, that I could not blow it above half full with the greatest expiration that I could make : And at the same time I could plainly perceive that my lungs were much fallen, just in the same manner as when we breathe out of them all the air we can at once. Whence it is plain that a considerable quantity of the elasticity of the air contained in my lungs, and in the bladder was destroyed : Which supposing it to be 20 cubic inches, it will be 1/13 part of the whole air, which I breathed to and fro ; for the bladder contained 74 cubic inches, and the lungs, by the following Experiment [Hales here refers to the flooding of the lungs of a calf with water to determine their volume] about 16 cubic inches, in all 240."

" These effects of respiration on the elasticity of the air, put me upon making an attempt to measure the inward surface of the lungs, which by a wonderful artifice are admirably contrived by the divine artificer, so as to make their inward surface to be commensurate to an expanse of air many times greater than the animal's body." Hales' estimate is about 289 square feet or about 32 square metres, which is about ten times the surface of a man's body.

In the subsequent discussion of this and other facts we shall follow Hales through a vacillation between the near grasping of the oxygen hypothesis, back to the elasticity hypothesis, and to and fro. In his commentary on the area of the lungs he comes near to the oxygen hypothesis, but remains in thrall to the idea of elasticity. He admits that his estimate may not be exact, " yet," he thinks, " it is evident from the foregoing experiments on respiration, that some of the elasticity of the air which is inspired is destroyed ; and that chiefly among the vesicles, where it is most loaded with vapours ; whence probably some of it, together with the acid spirits, with which the air abounds, are conveyed to the blood, which we see is by an admirable contrivance there spread into a vast expanse, commensurate to a very large surface of air, from which it is parted by very thin partitions ; so very thin as thereby probably to admit the blood and air particles (which are there continually changing from an elastic to a strongly attracting state) within the reach of each other's attraction, whereby a continued succession of fresh air may be absorbed by the blood. . . . Yet when we consider how much air continually loses its elasticity in the lungs, which seem purposely framed into innumerable minute meanders, that they may thereby the better seize, and bind that volatile *Hermes* : It makes it very probable, that those particles which are now changed from an elastic repulsive, to a strongly attracting state, may easily be attracted through the thin partition of the vesicles, by the sulphureous particles which abound in the blood."

Assured of the plausability of his elasticity hypothesis from this and other considerations Hales now made strenuous and ingenious efforts to falsify the oxygen hypothesis directly in confrontation with the elasticity hypothesis. The explanation of the effect of bad air on the lungs " has hitherto been supposed to be wholly owing to the loss and waste of the *vivifying spirit of air* ; but not unreasonably be also attributed to the loss of a considerable part of the air's elasticity, and the grossness and density of the vapours, which the air is charged with ; for mutually attracting particles, when floating in so thin a medium as the air, will readily coalesce into grosser combinations : Which effect of these vapours, having not been duly observed before, it was concluded, that they did not affect the air's elasticity ; and that consequently the lungs must needs be as much dilated in inspiration by this, as by a clear air ".

Hales believed that by mechanically restoring the " elasticity of the air " he would also restore its vivifying power. So he tried to do this, and supposed himself successful. His experiment used a live dog, whose respiration was maintained by an air pump consisting of a bladder which could be compressed. The dog showed distress, breathing in and out from the bladder, as Hales himself had felt when doing the same thing. But then, when this effect occured, he forced the air into the lungs of the dog by compressing the bladder and so mechnically restoring the *elasticity* of the air and lungs. The explanation, he thought, of why the dog continued to live a whole hour under this treatment, when a candle would have gone out long since in the amount of air in the bladder, must be due " to the forcible dilatation of the lungs, by compressing the bladder, and not to the *vivifying spirit of air* " as John Mayow would certainly have supposed. A fatal flaw is easily found in the experiment. It was frequently disturbed, Hales says, " by being obliged to blow more air into the bladder twelve times during the hour ! " Of course just enough oxygen was thus admitted to maintain the life of the miserable dog. The goodness of fresh air for respiration was put down by Hales to the way vapours " do in some measure clog and lower the air's elasticity ; which it best regains by having these vapours dispelled by the ventilating motion of the free open air, which is rendered wholesome by the agitation of the winds ". And all his reforming notions on air, which had such dramatic practical application in his ventilators, were based upon this very erroneous notion.

Plants, too, destroyed the elasticity of the air, by imbibing elastic air. In May he " set some well rooted plants of spearmint in two glass cisterns full of water, which cisterns were set on pedestals, and had inverted chemical receivers put over them, as in (fig. 16) the water being drawn up to *a a*, half way their necks : In this enclosed moist state the plants looked pretty florid for a month, and made, as I think, some few weak lateral shoots, though they did not grow in height ; they were not quite dead after six weeks, when it was found that the water had risen in both glasses from *a a* towards *z z*, in bulk about 20 cubic inches ". Hales was unable to cite this experiment with confidence since he had omitted to take the temperature at the beginning and the end of the experiment and was unable to say how much of the contraction in the air under the chemical receiver was due to the weather being cooler and how much to absorption by the mint. He suggests that the experiment should be done again with thermometers hung up to keep a check on the temperature, " and for greater certainty, it would be advisable to suspend in the same manner another like receiver with no Mint, but only water in it, up to *a a* ". This precaution we now call a " control experiment ". Whatever difference there is between the control and the actual experiment must be due to the processes in the actual experiment. In this way the most complex conditions can be taken account of without the immense difficulties of making corrections to the original experiment.

The elastic theory of the air and its role in vital processes in virtue of that elasticity are a wholly unfamiliar theory to us today. Yet it was of considerable influence in its time. Hales sums it up as follows : " Thus upon the whole, we see that air abounds in animal, vegetable and mineral substances ; in all which it bears a considerable part : if all the parts of matter were only endued with a strongly attracting power, whole nature would then immediately become one unactive cohering lump ; wherefore it was absolutely necessary, in order to the actuating and enlivening this vast mass of attracting matter, that there should be everywhere inter-mixed with it a due proportion of strongly repelling elastic particles, which might enliven the whole mass, by the incessant action between them and the attracting particles : And since these elastic particles are continually in great abundance reduced by the power of the strong attracters, from an elastic to a fixed state ; it was therefore necessary that these particles should be endued with a property of resuming their elastic state, whenever they were disengaged from that mass, in which they were fixed ; that thereby this beautiful frame of things might be maintained, in a continual round of the production and dissolution of animal and vegetable bodies.

The air is very instrumental in the production and growth of animals and vegetables, both by invigorating their several juices while in an elastic state, and also by greatly contributing in a fixed state to the union and firm connection of the several constituent parts of these bodies, *viz* their water, salt, sulphur and earth. This band of union, in conjunction with the external air, is also a very powerful agent in the dissolution and corruption of the same bodies, for it makes one in every fermenting mixture; the action and re-action of the aereal and sulphureous particles is in many fermenting mixtures so great as to excite a burning heat, and in others a sudden flame: And it is we see by the like action and re-action of the same principles, in fuel and the ambient air, that common culinary fires are produced and maintained".

Further reading

The final discovery of oxygen and the subsequent correct interpretation of many of these experiments did not take place until the end of the eighteenth century. In the years 1774 and 1775 Priestley did the essential experiments, but again they were misinterpreted. See Charles Singer, *A Short History of Scientific Ideas*: pp. 337–8. It was not until 1784 that Lavoisier finally gave the correct account (cf. Singer, pp. 339–42), the same account in essentials as Mayow had given in 1670.

Practical work

1. The experiment on the absorption of air through the bark can be easily repeated, following Hales' account. To avoid problems about sealing the added tube for the experiment on the effect of water immersion, it is sufficient simply to soak the length of stick in water, and show that in a thoroughly soaked condition air is not nearly so freely drawn through the piece of branch.

2. Using the usual chemical apparatus for preparing gases the following "distillations and fermentations" can be performed:

i. 10 gramme of ordinary meat can be put into a Pyrex flask and the gas evolved during heating collected. The meat should be reduced to "scoria", that is, completely charred.

ii. Ordinary oatmeal provides an adequate substitute for Indian Wheat, and can similarly be heated to destruction.

iii. Similarly the gas can be collected from the "fermentation" of oystershell (or eggshell) and vinegar. Why do you think that, on leaving the apparatus to stand, the "air is resorbed"? What gas do you think has been produced in this experiment?

iv. Either raisins in water, or yeast and sugar in water, will start to ferment. Collect the gas and again see if it is mostly resorbed after some days. Do you think this the same gas as is produced from eggshells or oystershells and vinegar?

3. Using a bell jar repeat Hales' experiment on burning paper soaked in sodium nitrate.

4. A mouse in a small cage placed under a bell jar, as in Hales' experiment on the rat, can be used to show that air is used up in breathing. The experiment is not easy to perform since the carbon dioxide

breathed out by the mouse may take some days to dissolve in the water so as to show the true loss of oxygen from the enclosed air. Perhaps this experiment is best read about, since there is a danger of suffocating the mouse should it be left under the jar a little too long !

5. A similar experiment can be performed with mint, as Hales suggests, but this time do not forget to take the temperature of the air, before and after the experiment, and to set up a control, as Hales himself ruefully recommends. Do you find that his conjecture as to the result of the proper performance of the experiment was correct ? In this experiment there is plenty of time for carbon dioxide expired by the plant to be dissolved in the water, so that the rise in the water gives a pretty accurate estimate of the amount of " elastic air " used up by the plant.

CHAPTER 10

vegetable life

A recurring theme in the discussions of the principles of biology has been the idea of *vitalism.* This is the theory that there is a special kind of force, or *perhaps a special* way of behaving, that is characteristic of living matter, and not found in the productions of inorganic nature. Opponents of vitalism have held that though plants and animals are very differently organized from inanimate things, nevertheless the principles by which they are organized, and in particular the forces and relations that bind together their ultimate parts, are the same as those in inorganic things. Hales was certainly not a vitalist. He believed that the structure of plants derived from the interaction of forces of attraction and repulsion between parts which are common to all of nature. He believed also that God has so arranged this in order to create plants, but that once His action as a designer had been complete, then ordinary natural forces sufficed, since God had created such " natural " forces, just so that they would suffice. His view comes out very strongly in his remarks on the cycle of life. He explains it thus : " but when the watery particles do again soak into and disunite the parts of plants and their repelling power is thereby become superior to their attracting power ; then is the union of the parts of vegetables thereby so thoroughly dissolved, that this state of putrefaction does by a wise order of Providence fit them to resuscitate again, in new vegetable productions ; whereby the nutritive fund of nature can never be exhausted : Which being the same both in animals and vegetables, it is thereby admirably fitted by a little alteration of its texture to nourish either ".

Hales had no new chemical theory to offer, and was happy to endorse the general view that plants were largely composed of sulphur, air, volatile salt, water and earth. Oil is a very characteristic and important plant substance, and is typical of the most highly refined parts of plants, its necessity in plant life being shown by the regular presence of oil in seeds. Oil, he held, was made of sulphur and air. This led him to a serious and retrograde error. We have already seen how Millington and Grew had clearly grasped the sexuality of plants, and thus the role of the pollen. Hales was seriously misled by the usually yellow colour of pollen into supposing that it was primarily sulphurous in nature, and that its function was to form a kind of cloud around plants and trees which would facilitate their attraction of suitable substances out of the air, of a sulphurous and airy nature, to enable them to build up their stock of oils. Not having grasped the idea that air is a mixture of gases, he was unable to conceive of carbon dioxide as a separate entity, and hence was very far from properly understanding the exchange of substances between the plant and the air.

Despite these biochemical handicaps he was able to form a mostly correct idea of the role of the leaves, though his views as to how the role was performed remained unsound. He assigns four functions to them, based ultimately upon their being the main seat of the passage of water into the air from the plant, and hence the source of the attractive power in other parts of the plant for water.

1. " The leaves are very serviceable," he says, " in this work of vegetation, by being instrumental in bringing nourishment from the lower parts, within reach of the attraction of the growing fruit; which like young animals is furnished with the proper instruments to suck it thence." Here again are the leaves exercising their power to draw sap towards them as the water evaporates off their surface. In this view Hales is wholly right.

2. " The leaves, in which are the main excretory ducts in vegetables, separate and carry off the redundant watery fluid, which by being long detained, would turn rancid and prejudicious to the plant, leaving the more nutritive parts to coalesce." Here again the role of the leaves in the perspiration of water serves the derivative role of excretion, thus drawing fresh sap upwards into the rest of the plant.

3. But not everything passes out of the leaves. Again Hales had hold of the right principle in thinking that " we may therefore reasonably conclude, that one great use of leaves is what has been long suspected by many, *viz.* to perform in some measure the same office for the support of the vegetable life, that the lungs of animals do, for the support of the animal life ; Plants very probably drawing through their leaves some part of their nourishment from the air ". This will happen, Hales correctly supposes, by the effects of changing external circumstances, such as the meteorological differences between night and day, which bring about changes in temperature, humidity and the like.

4. Finally Hales conjectures that light may also have a role, through the leaves, in the life of the plant. " And may not light also," he says, " by freely entering the expanded surfaces of leaves and flowers, contribute much to the en-nobling the principles of vegetables ? " That is to say light, he supposes, may have a role in the production of the more elaborate plant substances out of the simpler constituents drawn in through the roots and abstracted from the air. Newton himself had said in Query 30 of his *Opticks* " Are not gross bodies and light convertible into one another ? and may not bodies receive much of their activity from the particles of light, which enter their composition ? The change of bodies into light, and of light into bodies, is very conformable to the course of nature, which seems delighted with transmutation." But we should hardly take light as a *constituent* of organic nature now, though we should regard it as a source of " activity ", holding, as we now do, that light is a form of energy.

In Hales' studies on growth we return to an investigation of more permanent value than his uncertain excursions into chemistry and biochemistry. These experiments are of great beauty. He studied the growth both of shoots (that is stems) and of leaves, comparing the former with the growth of bones. He describes his investigations on shoots in Experiment CXXIII. " In order to find out the manner of the growth of young shoots, I first prepared the following instrument, *viz.* I took a small stick *a*, (fig. 20) and at a quarter of an inch distance from each other, I run the points of five pins, 1, 2, 3, 4, 5, through the stick, so far as to stand $\frac{1}{4}$ of an inch from the stick, then bending down the great end of the pins, I bound them all fast with waxed thread ; I provided also some red lead mixed with oil."

" In the spring, when the Vines had made some short shoots, I dipped the points of the pins in the paint, and then pricked the young shoots of the Vine, (fig. 20) with the five points at once, from *t* to *p* : I then took off the marking instrument, and placing the lowest point of it in the hole *p*, the uppermost mark, I again pricked fresh holes from *p* to *l*, and then marked the two other points *i h* ; thus the whole shoot was marked every $\frac{1}{4}$ inch, the red paint making every point remain visible."

Figure 22 shows the true proportion of the same shoot, when it was full grown, the *September* following ; where every corresponding point is noted with the same letter.

" The distance from *t* to *s* was not enlarged above $\frac{1}{60}$ part of an inch ; from *s* to *q*, the $\frac{1}{26}$ of an inch ; from *q* to *p*, $\frac{3}{8}$; from *o* to *n*, $\frac{6}{10}$; from *n* to *m* $\frac{9}{10}$, from *m* to *l*, $1+\frac{1}{10}$ of an inch ; from *l* to *i*, $1+\frac{3}{10}$ inch nearly ; from *i* to *h* three inches nearly." Thus the growth in length of shoots is not even throughout the length of the shoot but takes place very much more vigorously towards the tip. Growth is brought about by the pressure of the sap which dilates the soft tissues, at the tip of the shoot, and which is prevented from expanding the shoot into a globe by lateral " bridges " across the shoot, " But a dilating spongy substance, by equally expanding itself every way, would not produce an oblong shoot, but rather a globose one, like an Apple ; to prevent which inconvenience we may observe, that nature has provided several Diaphrams, besides those at each knot, which are placed at small distances across the pith ; thereby preventing its too great lateral dilatation."

Unlike shoots, bones grow at both ends, but not in the middle and this Hales showed by a very fine experiment. " And as in vegetables, so doubtless in animals, the tender ductile bones of young animals are gradually increased in every part, that is not hardened and ossified ; but since it was inconsistent with the motion of the joints to have the ends of the bones soft and ductile as in vegetables ; therefore nature makes a wonderful provision for this at the glutinous serrated joining of the heads to the shanks of the bones ; which joining while it continues ductile the animal grows, but when it ossifies then the animal can no longer grow. As I was assured by the following experiment, *viz.*

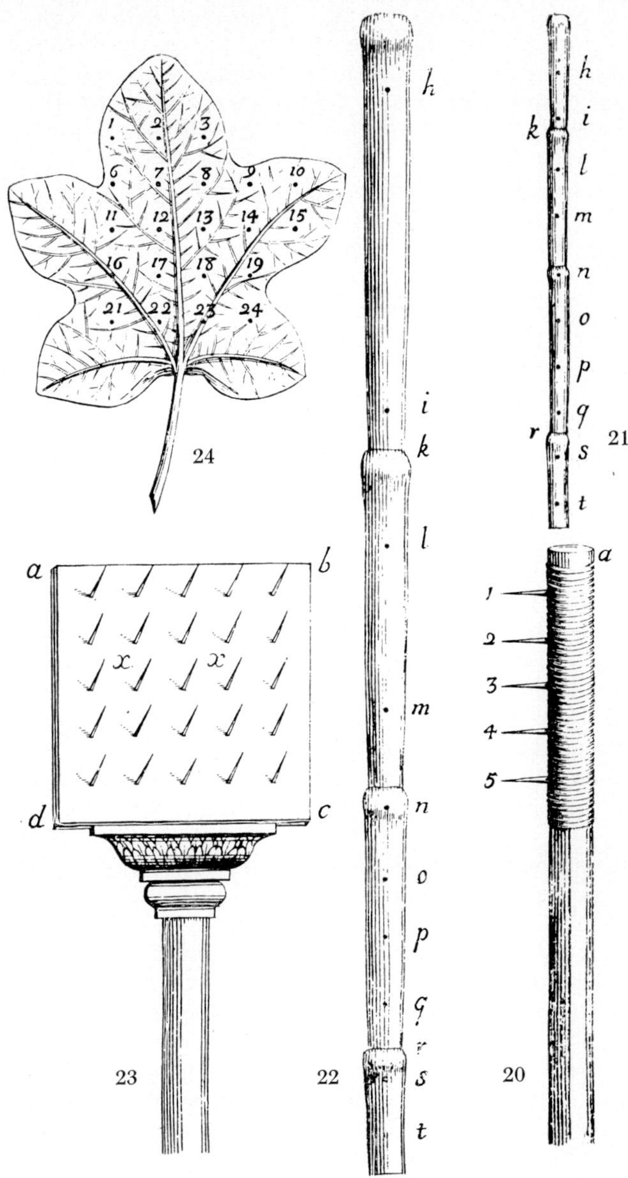

Figs. 20–24. Figs. 21 and 22 show the growth of a shoot by observing the markings made by the use of the instrument illustrated in fig. 20.

I took a half grown Chick, whose leg-bone was then two inches long, and with a sharp pointed Iron at half an inch distance I pierced two small holes through the scaly covering of the leg, and shin-bone ; two months after I killed the Chick, and upon laying the bones bare, I found on it obscure remains of the two marks I had made at the same distance of half an inch : So that that part of the bone had not at all distended lengthwise, since the time that I marked it : Notwithstanding the bone was in that time grown an inch more in length, which growth was mostly at the upper end of the bone, where a wonderful provision is made for its growth at the joining of its head to the shank, called by the Anatomists, Symphysis."

The growth of leaves can be studied in a somewhat similar way to the growth of shoots. "In order to enquire into the manner of the expansion of leaves," says Hales, " I provided a little Oaken board or spatula, *a b c d* of this shape and size (fig. 23) through the broad part at a quarter of an inch distance from each other ; I run the points of 25 pins *x x* which stood $\frac{1}{4}$ inch through, and divided a square inch into 16 equal parts."

" With this instrument in the proper season, when leaves were very young, I pricked several of them through at once, with the points of all these pins, dipping them first in the red lead, which made lasting marks.

Figure 24 represents the shape and size of a young Fig-leaf, when first marked with red points, $\frac{1}{4}$ inch distance from each other.

Figure 25 represents the same full grown leaf, and the numbers answer to the corresponding numbers in the young leaf : Whereby may be seen how the several points of the growing leaf were separated from each other, and in what proportion, *viz.* from a quarter of an inch, to about three quarter's of an inch distance.

In this Experiment we may observe that the growth and expansion of the leaves is owing to the dilatation of the vescicles of every part, as the growth of the young shoot was shown to be owing to the same cause in the foregoing Experiment ; and doubtless the case is the same in all fruits."

The plant is envisaged by Hales as being enlarged by the pressure of sap, which is not pushed up from below, but drawn upwards by the attracting power of the leaves. Thus he is enabled to conceive of the whole complex structure of a tree in these terms. " We may look upon a tree," he says, " as a complicated Engine which has as many different powers as it has arms and branches, each drawing from their common fountain of life, the root : And the whole of each yearly growth of the tree will be proportionable to the sum of their attracting powers, and the quantity of nourishment the root affords : But this attracting power and nourishment will be more or less, according to the different ages of the tree, and the more or less kindly seasons of the year."

" And the proportional growth of their lateral and top branches, in relation to each other, will much depend on the difference of their several attracting powers. If the perspiration and attraction of the lateral branches is little or nothing, as in woods and groves, then the top branches will mightily prevail ; but when in a free open air, the perspiration and attraction of the lateral branches comes nearer to an equality with that of the top, then are the aspirings of the top branches greatly checked."

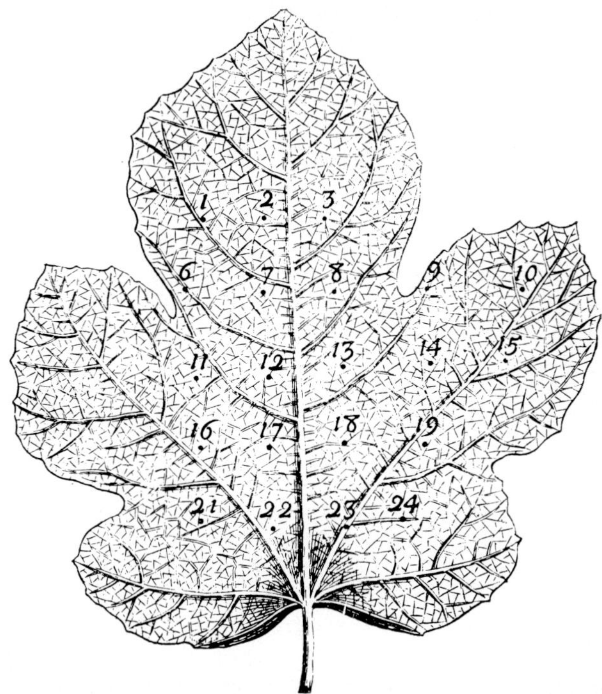

Fig. 25. The growth of a fig leaf. Fig. 23 shows the instrument for marking the young leaf (fig. 24).

Though we have omitted many details and many small discoveries we have now covered most of Hales' great work on the physiology of plants, and have followed him in his thoughts and his discoveries. The next one hundred years saw only very slow development in the study of plants. In fact *less* was commonly known in 1830 than was known in 1730. The next great burst of discovery came in the middle years of the nineteenth century. However, one or two discoveries of importance were made in the intervening time, and built more or less directly on Hales' work and by the use of his methods. They derived mostly from a much clearer idea of the chemistry of gases, which we have seen to have been fearfully muddled in Hales' mind. After 1780 Ingenhousz rediscovered John Mayow's ideas on respiration of plants, and showed quite clearly that there are two distinct respiratory processes, in one of which carbon dioxide is absorbed and oxygen eliminated, an essentially nutritive process, and in the other of which oxygen is absorbed and carbon dioxide eliminated, an essential respiratory process. Hales had known about both of these processes but had not had a clear enough idea of gases to make sense of the facts he had himself ascertained, nor could he understand Mayow's discoveries. Further investigation of these gaseous interchange processes was made by de Saussure. The final step in sorting out the gas exhange between plants and the air came in Boussingault's discovery, sometime after 1840, that the nitrogen in plants is not derived from the air, but from " fixed " nitrogen which is present in the earth, and is taken up by the roots. The discovery of the role of chlorophyll and clear ideas about the part light plays in the life of the plant had to wait until later in the century.

Further reading

In Hales' *Vegetable Staticks*, Chapter VII sums up Hales' ideas on the life of vegetables. A good description of developments after Hales can be found in J. Von Sach's *History of Botany*, translated by H. E. F. Garnsey and I. B. Balfour. Finally a good modern botany or plant physiology book should be consulted to get an idea of the present state of knowledge of the life of plants.

Practical work

1. The growth of shoots can be studied in just the way Hales used. It is probably easier to use a piece of hardboard and a line of five small nails, set 1 cm apart, and protruding about 1 cm through the board. The experiment can be performed on any garden plant which grows rapidly in early summer ; for example, the way the shoots of the rose grow can be easily determined. After marking the shoot first when it is about 10 cm long, it can be measured weekly and the *rate* of expansion of each 1 cm length discovered.

2. To study the growth of leaves a similar instrument to Hales' " spatula " can be made out of hardboard, spacing the small nails at the corners of 1 cm squares.

3. A comparison of the ratios of height to breadth of trees can be made, comparing two specimens of the same species, of about the same age, one growing in the open air, and the other in a grove. Do their comparative dimensions bear out Hales' idea that in the free air the lateral branches can perspire more freely and so draw more sap for growth in breadth ?

INDEX

THE WYKEHAM SCIENCE SERIES

for schools and universities

1 *Elementary Science of Metals* — J. W. Martin and R. A. Hull
(S.B. No. 85109 010 9) — **20s.—£1.00 net** *in U.K. only*

2 *Neutron Physics* — G. E. Bacon and G. R. Noakes
(S.B. No. 85109 020 6) — **20s.—£1.00 net** *in U.K. only*

3 *Essentials of Meteorology* — D. H. McIntosh, A. S. Thom and V. T. Saunders
(S.B. No. 85109 040 0) — **20s.—£1.00 net** *in U.K. only*

4 *Nuclear Fusion* — H. R. Hulme and A. McB. Collieu
(S.B. No. 85109 050 8) — **20s.—£1.00 net** *in U.K. only*

5 *Water Waves* — N. F. Barber and G. Ghey
(S.B. No. 85109 060 5) — **20s.—£1.00 net** *in U.K. only*

6 *Gravity and the Earth* — A. H. Cook and V. T. Saunders
(S.B. No. 85109 070 2) — **20s.—£1.00 net** *in U.K. only*

7 *Relativity and High Energy Physics* — W. G. V. Rosser and R. K. McCulloch
(S.B. No. 85109 080 X) — **20s.—£1.00 net** *in U.K. only*

8 *The Method of Science* — R. Harré and D. Eastwood
(I.S.B.N. 0 85109 090 7) — **25s.—£1.25 net** *in U.K. only*

9 *Introduction to Polymer Science* — L. R. G. Treloar and W. F. Archenhold
(I.S.B.N. 0 85109 100 8) — **30s.—£1.5 net** *in U.K. only*

10 *The Stars: their structure and evolution* — R. J. Tayler and A. S. Everest
(I.S.B.N. 0 85109 110 5) — **30s.—£1.5 net** *in U.K. only*

11 *Superconductivity* — A. W. D. Taylor and G. R. Noakes
(I.S.B.N. 0 85109 120 2) — **25s.—£1.25 net** *in U.K. only*

(*Price for* 8–11 *to be announced later*)

THE WYKEHAM TECHNOLOGICAL SERIES

for universities and institutes of technology

1 *Frequency Conversion* — J. Thomson, W. E. Turk and M. Beesley
(S.B. No. 85109 030 3)

2 *The Art and Science of Electrical Measuring Instruments* — E. Handscombe
(I.S.B.N. 0 85109 130 X)

Price per book for the Technological Series **25s.—£1.25 net** *in U.K. only*

NOTES